John Gribbin is a Visiting Fellow in Astronomy at the University of Sussex, and author of many books including *Science: A History*. He has long had a special interest in the weather, and is a Fellow of the Royal Meteorological Society.

Mary Gribbin studied psychology and now works in education. Her books include *Ice Age* and, for younger readers, *Big Numbers*. She has a special interest in exploration and is a Fellow of the Royal Geographical Society.

FitzRoy

The Remarkable Story of Darwin's Captain and the Invention of the Weather Forecast

John and Mary Gribbin

review

First published in the UK in 2003
by REVIEW

An imprint of Headline Book Publishing

First published in paperback in 2004
by REVIEW

10 9 8 7 6 5 4 3 2 1

Cataloguing in Publication Data is available from the British Library.

ISBN 0 7553 1182 5

Typeset in Caslon by Avon DataSet Ltd,
Bidford-on-Avon, Warwickshire

Printed and bound in Great Britain by
Mackays of Chatham PLC, Chatham, Kent

Headline's policy is to use papers that are natural, renewable and recyclable
products made from wood grown in sustainable forests. The logging
and manufacturing processes are expected to conform to the
environmental regulations of the country of origin.

HEADLINE BOOK PUBLISHING
A division of Hodder Headline
338 Euston Road
London NW1 3BH

www.reviewbooks.co.uk
www.hodderheadline.com

For Charlie Munger, with thanks

Contents

ACKNOWLEDGEMENTS

We are grateful to the following people and institutions for help and advice:

In England, the Admiralty Library, Sir Thomas Barlow, British Library Reference Service, Jim Burton, Centre for Kentish Studies, Dorset Record Office, Durham County Record Office, the Hydrographic Department of the Ministry of Defence, Richard Keynes, Simon Keynes, the Meteorological Office Library and Archive, Public Record Office at Kew, Royal Geographical Society, Royal National Lifeboat Institution, Royal Naval Museum and Library, Royal Society, University of Cambridge Library, University of Sussex Library.

In New Zealand, Hilda and Graham Heap and Fiona Donald scoured the archives for FitzRoy papers on our behalf. Thanks also to the Alexander Turnbull Library and National Library of New Zealand.

In Australia, Leo van de Pas, the Dixson Library, State Library of New South Wales, and the La Trobe Library, State Library of Victoria.

And of course, Tim Berners-Lee, Apple computers, and Google.

The Alfred C. Munger Foundation contributed to our travel and research expenses.

Chapter Four draws on material from our book *Darwin in 90 Minutes*, which is now out of print.

We follow the convention of using italics for emphasis in quotations wherever the original handwritten documents use underlining.

Those who never run any risk; who sail only when the wind is fair; who heave to when approaching land, though perhaps a day's sail distant; and who even delay the performance of urgent duties until they can be done easily and quite safely; are, doubtless, extremely prudent persons: – but rather unlike those officers whose names will never be forgotten while England has a navy.

Robert FitzRoy

He is an extra ordinary but noble character, unfortunately however affected with strong peculiarities of temper. Of this, no man is more aware than himself, as he shows by his attempts to conquer them. I often doubt what will be his end, under many circumstances I am sure it would be a brilliant one, under others I fear a very unhappy one.

Charles Darwin

INTRODUCTION

A Very British Hero

CHARLES DARWIN SUMMED up the character of his Captain on the famous voyage of the *Beagle* in these words:

As far as I can judge, he is a very extraordinary person. I never before came across a man I could fancy being a Napoleon or a Nelson. I should not call him clever, yet I feel convinced nothing is too great or too high for him. His ascendancy over everybody is quite curious: the extent to which every officer and man feels the slightest rebuke or praise, would have been before seeing him, incomprehensible ... His greatest fault as a companion is his austere silence: produced from excessive thinking: his many good qualities are great and numerous: altogether he is the strongest marked character I ever fell in with.

Yet the name of this man, who could have been a Napoleon or a Nelson, and who, apart from his achievements with the *Beagle* (which far exceed what you might guess from reading biographies of Darwin), was in fact also Governor of New Zealand and as founder of the UK Meteorological Office the world's first full-time professional weather forecaster, remains known to most people only through the association with Darwin. We hope this book will redress the balance, presenting the true story of a very British hero of the nineteenth century, whose career began and ended with tragedy.

As his surname suggests, FitzRoy came from aristocratic stock; he was a direct descendant of the first Duke of Grafton, the formally acknowledged illegitimate son of Charles II, resulting from his notorious liaison with Barbara Villiers, the Duchess of Cleveland. He first made his name in the Royal Navy when he was given command of the *Beagle* when the Captain, partly as a result of the loneliness of command under the rigid hierarchy of the Navy in those days, became depressed and shot himself. It was partly because of fear that he might go the same way himself that on his second voyage in command of the *Beagle* FitzRoy took along a gentleman companion, someone he could talk to as an equal – a certain Charles Darwin.

FitzRoy was born in 1805, the year Nelson died at Trafalgar; his background, early life and career in the post-Nelsonian Royal Navy, the voyage with Darwin and its aftermath, would alone make for a fascinating story, full of the excitement and adventure of life under sail, in the mould of the novels of Patrick O'Brian. He was a superb seaman, navigator and surveyor, almost in the mould of Captain James Cook or (for all his faults) Captain William Bligh. It is hardly FitzRoy's fault that he was born too late to discover Australia, but the work he did on two surveying voyages along both the Atlantic and the Pacific coasts of South America, still in the days of sail, ranks with the best of its kind. Spice is added to the mix by the fact that FitzRoy, already a devout Christian, became something of a fundamentalist after he married a deeply religious lady, and after the publication of Darwin's *Origin of Species* in 1859 he became a bitter and outspoken opponent of the theory of evolution by natural selection.

Before that, however, FitzRoy had served in Parliament for two years as a Tory MP, being involved in a scandal which almost led to a duel, before being appointed in 1843 as Governor of New Zealand, a post in which he quickly made himself unpopular with the British colonists and the powers-that-be at home by trying to uphold Maori rights in a fair manner (almost uniquely for anyone in such a position at that time, he believed that treaties signed with the natives should be upheld). He was recalled after a brief term of office, without even receiving the customary knighthood. But to modern eyes his failings seem eminently reasonable – he failed to crack down hard on the Maori population, and in trying to be fair to everyone managed to offend the white settlers. Recalled to Britain in 1845, he returned to active naval service, where he commanded the first screw-driven ship to be commissioned by the Royal Navy, and developed his interest in meteorology.

FitzRoy was interested in the weather for one reason – to save lives. He knew from direct experience the value of advance warning of storms at sea, and was determined to do something to help his fellow mariners. This was an outstanding example of his sense of duty, a noblesse oblige of the best kind which drove him to spend his own fortune in government service, leaving only debts for his wife and children, to do what he thought right for the common good at all times regardless of the effect on his own reputation, and to work long hours that far exceeded his formal obligations. In the words of one of his obituaries, 'a more high-principled officer, a more amiable man, or a person of more useful general attainments never walked a quarter-deck'.

It is a sign of FitzRoy's strength of character that even after the setback in New Zealand, back in England he developed the fundamental techniques of weather forecasting, designed a standard barometer and thermometer (a prototype weather station), invented the system of storm warnings and signals which saved countless lives in the ensuing decades, and issued the first daily weather forecasts, published in *The Times* – indeed, he invented the term 'weather forecast'.

If it had not been for Robert FitzRoy, the name Charles Darwin would now be remembered, if at all, as that of a country parson with an interest in natural history, perhaps rather in the mould of Gilbert White, of Selborne. The theory of natural selection, which explains the fact of evolution, would be known from the work of Alfred Russel Wallace, who came up with the idea independently of Darwin, and whose work prompted Darwin to go public with his own ideas; we would be as familiar then with the term 'Wallacian evolution' as we are, in the real world where Robert FitzRoy lived, with the term 'Darwinian evolution'. In that real world, FitzRoy is known, so far as he is widely known at all, as Darwin's Captain on the voyage of HMS *Beagle* during which the young naturalist made the observations which provided the inspiration for the further years of hard work on which his theory would be based. But if Charles Darwin had never lived, the name of Robert FitzRoy[1] might be widely held in higher esteem than it is in our world, where it has remained forever in the shadow of Darwin.

By any ordinary standards, Robert FitzRoy achieved much in a life where the successes far outweighed the failures; but in his own mind he failed to live up to the standards he set himself (and others). He was also highly sensitive to criticism, depressive, and frustrated by the failure of nineteenth-century technology to provide the reliable weather forecasts that he knew were feasible in principle and which would be so valuable to society at large, not just to sailors. Robert FitzRoy never did get the knighthood he so richly deserved – but then, neither did Charles Darwin. Whatever his achievements, it is inevitable that his name will always be linked with that of his most famous passenger; but we hope that our biography will do something to bring him out from Darwin's shadow.

John Gribbin
Mary Gribbin
December 2002

CHAPTER ONE

Before the *Beagle*

ROBERT FITZROY WAS a member of an aristocratic English family that began with the liaison between King Charles II and Barbara Villiers in the second half of the seventeenth century.[1] Henry, the illegitimate son resulting from that liaison, born in 1663, was made the first Duke of Grafton; the family name FitzRoy is the traditional one for an acknowledged royal bastard, from the Norman French son (*fils*) of the king (*le roi*). Brought up in the hothouse atmosphere of the Restoration, but widely acknowledged as 'the ablest and best of all Charles's acknowledged sons',[2] the young Duke served his father as an Admiral on board his own ship, the *Grafton* (apparently ably, although this was clearly an appointment which owed nothing to his ability as a sailor) as well as in the Army. He then saw his uncle James, a Catholic, succeed Charles in 1685, as James II of England and VII of Scotland. At first, Henry served James equally loyally, playing a part in the crushing of the Monmouth Rebellion of 1685. But he shared with others the growing unease with James's uncompromising Catholicism. In 1688, James was ousted, at the instigation of Parliament, by the Protestant William of Orange and his wife Mary, the daughter of James II and first cousin of Lord Grafton. Henry played a significant role in what has gone down in history as 'the Glorious Revolution', switching sides, together with John Churchill (later the first Duke of Marlborough), out of what seems to have been a genuine desire for the good of the country, rather than opportunism. Just two years later, Henry was killed in action against supporters of James in a battle near Cork. By then, however, the twenty-seven-year-old, who had married young, already had a seven-year-old son.

His heir Charles, the second Duke (1683–1757), had a less tempestuous life, and became Lord Chamberlain to both George I and George II, acting as head of the royal household and running all aspects of its day-to-day management.[3] The Lord Chamberlain was at that time an ex officio member of the Cabinet, and so Grafton was very much at the centre of political life in England for the thirty-three years he held the post, but he never got his hands on the levers of power. One of the other duties of the Lord Chamberlain, which he retained (unlike his Cabinet post) until into the second half of the twentieth century, was to act as censor for plays performed in England. This may partly account for the bitterness of Jonathan Swift's description of

> Grafton the deep,
> Either drunk or asleep.

Charles did not marry until he was thirty, in 1713. His first wife bore him three sons before dying in childbirth in 1726 at the age of thirty-six, but none of them survived to inherit the title. This was just as well in the case of the eldest, George, who was born in 1715 and grew up to be a vicious brute. He married the beautiful seventeen-year-old Lady Dorothy Boyle in 1741. Seven months later, the pregnant girl was dead, officially of smallpox but in fact as a result of her husband's 'extreme brutality, the details of which are almost too revolting to be believed'.[4] The third son of the second Duke, through whom the title would pass, was Lord Augustus FitzRoy, who seems to have inherited more of the temperament of the first Duke. He joined the Navy and was serving in American waters when he met his future wife, the daughter of the Governor of New York. Without bothering to obtain the Duke's permission, they married at once, in 1733, although Lord Augustus was only seventeen. The Duke accepted the fait accompli with good grace and was delighted with the arrival of his grandson, Augustus Henry FitzRoy, in 1735. Lord Augustus remained in the Navy and was appointed to command his own ship at the age of twenty-one, but he died of fever at Jamaica in 1741. The disgusting Lord George died in 1747, by which time the middle brother had also died. With the widow of Lord Augustus remarrying, the heir to the title, Augustus Henry, was brought up with his younger brother Charles in the household of the old Duke, where he was groomed for a life in politics. He succeeded to the title as the third Duke of Grafton in 1757, at the age of twenty-two. By then, he had been educated at Cambridge, taken the gentlemanly Grand Tour of the continent, been elected MP for Bury St Edmunds, and married an heiress, Anne Liddell.

The third Duke is sometimes portrayed as a man who enjoyed the quiet life and had few ambitions. For such an unambitious aristocrat in eighteenth-century England, politics certainly made a pleasant pastime. But as a member of the Whig party Grafton rose to become First Lord of the Treasury in 1766, at the age of thirty-one, and then an accidental Prime Minister when William Pitt (the Elder) was taken ill in 1767 and someone had to step in to the breach (so, in a sense, the Stuarts did regain power in Britain!)[5]. Even in eighteenth-century Britain, though, an aristocrat did not become First Lord of the Treasury and Prime Minister without some ability and a capacity for hard work, and this Grafton, like many of his line, seems to have had a strong sense of duty which for a long time kept him hard at work in the profession chosen for him by his grandfather, even if he might not have chosen it himself. But this is not the place to go into the politics of the time. Suffice it to say that his term as acting Prime Minister was not exactly a success, and in 1770 his coalition government was replaced by a Tory administration headed by Lord North and openly supported by the King,

George III. It was Lord North who, backed by the King, took the hard line with the American colonies which led to the War of Independence. Pitt (and Grafton) had supported the claims of the colonists, but Grafton's weakness as Prime Minister was one of the factors that left the way open for the hardliners, and thereby lost Britain the American colonies.

The third Duke lived until 1811, enjoying a reasonably quiet later life (earlier excitements included a long relationship with a mistress, divorce in 1769 and a second marriage, not to the mistress, that same year).[6] The first wife of the third Duke had given him just one daughter; his second Duchess, twenty-three at the time of their marriage to his thirty-four, produced twelve children, and the Duke lived to see the arrival of several grandsons, including Robert FitzRoy in 1805. Robert was the son of one of the younger sons of the third Duke, Lord Charles FitzRoy; the 'Lord' is a courtesy title given to younger sons of the nobility, but the courtesy does not extend to later generations, so Robert FitzRoy, although undoubtedly an aristocrat, was never a Lord, although he was directly descended from a King, several Dukes and a Prime Minister. The fourth Duke, Robert's uncle, followed the example of the third Duke's later life, as a country gentleman, famous only for owning a horse that won the Derby (Whisker, in 1815).[7] Another uncle, Lord William FitzRoy, joined the Navy and rose to become an Admiral. Lord Charles studied at Cambridge University, joined the Army and rose to become a General. In later life he entered Parliament (we use the term loosely), and served (?) Bury St Edmunds for a quarter of a century without ever speaking in the House. Like his brother, the Duke, he devoted his time to the traditional pursuits of a country gentleman. These included marriage and raising a family, but his first wife, by whom he had one son (Charles, born in 1796), died, and in 1799 Lord Charles married for a second time, to Lady Frances Anne Stewart (known as Fanny to her family). She was a member of an aristocratic family with a more recent, and slightly less exalted, pedigree than that of the FitzRoys, but which included a man who would become one of the most important political figures in Europe (perhaps *the* most important, in the aftermath of the Napoleonic Wars), Robert Stewart, Lord Castlereagh, the half-brother of Lady Frances.

The Stewarts were an Irish family of Scottish stock. The family had settled in Ireland, in the county of Donegal, early in the seventeenth century, but their rise to political prominence really began more than a century later, when Alexander Stewart, a successful Belfast businessman, succeeded his elder brother to the family estate and married (in 1737) an heiress, Mary Cowan, whose money he used to buy two more estates, in County Down. Alexander's own political ambitions came to nothing, but in two succeeding generations the power and influence of the wealthy land-owning family grew. Alexander's eldest son, Robert, was born in 1739, and in 1766 he married Lady Sarah Seymour-Conway, the daughter of the Lord Lieutenant of Ireland (the Earl of Hertford), at a ceremony held in the Chapel Royal of Dublin Castle. But the marriage which started with such a flourish was to be short and marred by tragedy. Their first son, also called Alexander, was born in 1768, and a second son, Robert, on 18 June 1769, in the same month that baby Alexander died. In 1770, during the late stages of her third pregnancy, Lady Sarah and her unborn child both died from complications associated with the pregnancy. Alexander married again in 1775, again choosing the daughter of an English aristocrat, this time Lady Frances Pratt, the eldest daughter of the Earl of Camden. Lady Frances had better luck than her predecessor; she produced eleven children, only one of whom died in infancy. The eldest of these, named Frances after her mother, was born in 1777, and became the mother of 'our' Robert FitzRoy (himself named in honour both of his maternal grandfather and of Fanny's half-brother) on 5 July 1805, at the family home of Ampton Hall, near Bury St Edmunds, in Suffolk. He was the third child and second son of the marriage between Fanny and Lord Charles FitzRoy.

But before we begin our story of the life of Robert FitzRoy, there is something more to be said about the life of Robert Stewart, his uncle, since the manner of his uncle's death was to cast a shadow over FitzRoy, and was a minor (perhaps not so minor) contribution to his decision to take a gentleman companion with him on his second voyage in command of the *Beagle*. Robert Stewart was born with the proverbial silver spoon in his mouth, a member of a rich land-owning family which was now connected by marriage to the English aristocracy. His father represented County Down in the Irish House of Commons until 1789, when he was created Baron Londonderry thanks to the influence of his father-in-law, Lord Camden. This made him ineligible to sit in the Irish House of Commons (but would not have prevented him, had he wished, from seeking election as an MP for the Parliament in London, since it was an Irish title). There was, however, an ideal candidate to take his place, in the form of the younger Robert Stewart, who had just spent two years in Cambridge (without taking a degree, not then regarded as an essential requirement for a gentleman), and who, after the expenditure of a great deal of his father's fortune, was duly elected in 1790 while still some weeks short of his twenty-first birthday, this legal technicality apparently counting for little compared with the influence of the Stewart family and its connections. Thereafter, the roles were soon reversed, and Baron Londonderry's successive elevations in the Irish peerage (to Viscount Castlereagh in 1795, Earl of Londonderry in 1796 and Marquess of Londonderry in 1816) were more in recognition of his son's achievements than of his own, it being anticipated that the son would duly inherit all these titles. It was when the elder Robert Stewart was elevated from Viscount Castlereagh to Earl of Londonderry in 1796 that the subsidiary title automatically became the courtesy title of the younger Robert Stewart, which is why he has gone down in the history books as Lord Castlereagh. But neither father nor son ever accepted an English title, since taking up such a title (as Castlereagh would have been forced to do when his father died, even if he had not originally accepted the title himself) would have prevented him from sitting in the House of Commons at Westminster, where he had become so essential to the government – but that is running ahead of our story.

At that time, Ireland had its own Parliament, which was technically independent of Westminster, but which in fact largely followed the 'advice' of the King's representative in Dublin, the Lord Lieutenant of Ireland. The American War of Independence, the writings of Tom Paine, and the revolution in France all helped to fan the flames of Irish independence, and in these difficult times Castlereagh (as we shall call him for consistency) soon showed genuine political skill, becoming indispensable to the administration. In March 1798 he was appointed

Acting Chief Secretary (in other words, right-hand man) to the Lord Lieutenant, the then Earl of Camden, Castlereagh's step-uncle. In spite of the family connection, this was no act of nepotism; he was so much the right man for the job that in spite of doubts about putting an Irish-born man in the post, in November that year he was formally awarded the post, no longer merely 'acting', by Camden's successor, Lord Cornwallis. Castlereagh was the only Irishman ever to hold the post, and played a key role in crushing the rebellion of 1798 as he had in beating off the threat of French invasion in 1797. Convinced by these threats that the only secure future for Ireland lay in union with Britain, he pushed the Act of Union through the Irish Parliament in 1800, for which his name is still reviled by many Irish people today. But such objectors miss the point that Castlereagh intended the union to be accompanied by the political emancipation of the Catholic majority population in Ireland, and that his efforts in this regard were only just thwarted by the opposition of George III. Catholics did not get the vote until 1829, under Wellington's Tory ministry.

With the Irish Parliament now even less of a meaningful body, Castlereagh sat in the House of Commons at Westminster, where as Minister for War he (among other things) reorganised the entire basis of the Army and militia, providing the fighting machine that Sir Arthur Wellesley (later the Duke of Wellington) would use so effectively against Napoleon's forces; he was also responsible for Wellesley's appointment as commander of that army in Spain. As Foreign Secretary and Leader of the House of Commons (posts he held simultaneously and continuously from 1812 until his death), Castlereagh was the main representative of the government in the Commons, since the Prime Minister, Lord Liverpool, sat in the House of Lords. He was also the face of Britain overseas, one of the greatest statesmen in Europe, instrumental in obtaining the peace treaty that followed the first defeat of Napoleon in 1814, and then, with the Austrian Klemens von Metternich, dominating the Congress of Vienna where the political future of Europe was decided after Napoleon's final defeat. It was Castlereagh, more than anyone, who established the policies and alliances which would keep Britain free from involvement in major continental wars until 1914.

But all this took its toll. Castlereagh worked under intense pressure even after the Napoleonic Wars, and by 1821 was showing the first signs of mental illness, possibly exacerbated by the death of his father in April that year. In 1822, this developed into full-blown paranoia, fuelled by his own belief (although nobody ever found evidence for this) that he was being blackmailed for alleged homosexual acts.[8] On 12 August 1822, a victim more than anything of decades of dedicated service to his country, Castlereagh cut his own throat. Perhaps the most significant tribute paid to him was uttered by a servant who was asked if he had noticed anything odd about his master just before his death. 'Yes,' the man replied, 'one day he spoke sharply to me.'[9] Even in a family as distinguished as the FitzRoys, having an uncle like that was a hard act to follow, and the manner of Castlereagh's death made a big impact on the seventeen-year-old Robert FitzRoy, then serving as a Midshipman in the Royal Navy.

Joining the Navy might have been a natural course for young Robert to follow in any case, with something of a tradition of naval service in the family (most recently embodied in his uncle, Lord William). But the usual practice of sending younger sons off to serve the King may have seemed even more attractive to Lord Charles following the death of Robert's mother in 1810, just a year after the family had moved to Wakefield Lodge, near the village of Pottersbury, in Northamptonshire. It was there that Robert lived with a widowed father, a half-brother eight years older than himself and an elder brother (George, born in 1800) and sister (Frances, born in 1803 and known as Fanny, like her mother), plus the inevitable retinue of servants. We have just one 'nautical' anecdote from young Robert's childhood, recounted in an unsigned obituary in *Good Words*.[10] Taking advantage of the servants' dinner hour, when he was unsupervised, the boy took a laundry tub, added a few bricks for ballast, and launched himself upon a large pond, successfully steering himself to the other side using a pole. It was only when he leaned over to stick his pole into the bank on the other side that he 'overbalanced the extemporised vessel; the bricks came sliding down, and in one moment, sailor, tub, and bricks were in the water. A gardener, who fortunately happened to be near, came to the rescue, and succeeded in drawing the young navigator to the bank, with no more injury than a thorough soaking.'

It wasn't long, though, before the boy was sent away to school. In 1811, when he was six, he was sent to school at Rottingdean, on the south coast of England near Brighton, and at the age of eleven he moved on to Harrow School, where he stayed for only a year before he was sent to the Royal Naval College at Portsmouth in February 1818, still five months short of his thirteenth birthday. Judging from the few surviving examples of his letters home, he seems to have been a serious, hard-working boy, always eager to seek the approval of his father. He doesn't seem to have joined in comfortably with the rough and tumble of school life, and his closest lifelong friend was his sister Fanny; his letters to her[11] give some of the few insights into his personal feelings that we get. The house he was brought up in (during the school holidays) was previously the hunting lodge of the Grafton family; but don't imagine this means it was a log cabin – rather, it was a Palladian mansion (albeit small as Palladian mansions go), built in the 1780s, located in a large private park with a lake big enough to sail on. In the days when country gentlemen really were country gentlemen, Lord Charles FitzRoy and his sons fitted the description to a T.

Apart from those few letters home, we have very little information about Robert FitzRoy's life at school and at home during those years, and scarcely any more about his own experiences during his early career in the Royal Navy. The evidence suggests that his education before joining the Naval College consisted of little more than the Classics and sport, since at the College he was initially placed in the lowest grade for everything except Classics. It soon became clear, though, that his ignorance was a result of bad (or non-existent) teaching, since he made rapid progress once he had the opportunity (and, indeed, received the College medal in mathematics at the end of his time there). On 20 May 1818 (still only twelve), he writes to tell his father that:

We had another Examination last Monday & I took[12] three more places which were all that I could take for I am now at the head of the part of the College in which I stay, so I could not have taken any more, I think I have got through pretty well as yet for I have not been reported either in or out of School & I have got a very good Character.

In another letter to his father, dated 9 August 1818, we get a glimpse of young FitzRoy's curiosity about the world around him:

> Last night when I was coming back the boat that I was in was going very quick, & I put my hand into the water & in the little ripple of water which it made Sparks, at least they looked exactly like it, kept coming from it like from a flint & steel & I could not make out what it was for the oars did the same a little, so I would be obliged to you if you will tell me when you write what it was, for I should like to know.

But these are rare glimpses into FitzRoy's personal world.

Although we have no insights into his personal life at this time, happily there is a wealth of information about what life was like in general for young gentlemen who chose (or had chosen for them) the route which FitzRoy followed into a naval career. A particularly apposite insight is provided by Bartholomew James Sulivan (usually known as James Sulivan), who followed the same route a few years later, and who served with FitzRoy before the *Beagle* voyages and while FitzRoy was in command of the *Beagle*. His autobiographical reminiscences, up to the point where he joined FitzRoy on the *Beagle*, form the early part of a life of Sulivan edited by his son.[13] A more general (and reasonably accurate) image of the life of a Midshipman in the Royal Navy of those days can be gleaned from the writings of C. S. Forester and Patrick O'Brian – and, of course, from the novels of Captain Marryat, the archetype and inspiration for the Hornblower and Aubrey novels.[14]

The one key point that does not come across from Forester and O'Brian is that it was not actually possible (according to the letter of the regulations) for a young gentleman to join a ship directly as a Midshipman. Without going into the complicated evolution of the chain of command and ranks of officers on board ships of the Royal Navy from the time of Samuel Pepys, whose work for the Navy Board and the Admiralty in the second half of the seventeenth century laid the foundations of the modern professional Royal Navy, by the end of the Napoleonic Wars there were two ways in which it was possible to become a Midshipman, gaining a foothold on the lowest rung of the ladder that might lead, through the ranks of Lieutenant, Commander and Captain, to the dizzy heights of Admiral. The traditional route, still the more common one in the early nineteenth century, was to start life at sea at a very young age as a volunteer, rated as an Able Bodied Seaman (AB), or as the servant to the Captain or one of the other ship's officers. Even here, there were originally two kinds of volunteers who entered a ship directly in this way – those approved by the Admiralty, who carried a letter of recommendation from the King and were known as 'King's Letter boys', and those who were the direct protégés of the Captain or another officer – possibly a relative, or the son of somebody the Captain owed a favour to. The idea was that they would learn seamanship and navigation on the job; eighteenth-century regulations stated clearly that no young gentleman could be rated Midshipman 'till they have served four years [at sea] and are in all respects qualified for it'. From 1731, they also stated that no young man should enter the service before the age of thirteen (eleven for the son of an officer). But both regulations were bent to the extreme in practice.

The first ploy was to enter a boy on the books of a ship long before he set foot on her, and even for him to be transferred (on paper) from ship to ship with the Captain who was his sponsor. There are documented instances of a child of one being entered in this way, and then being rated Midshipman as soon as he actually set foot on his ship, twelve or thirteen years later, since on paper he had served four years at sea already, even if nobody in their right mind could possibly claim that he was 'in all respects qualified' for the rank. There must have been many disastrous failures resulting from this cavalier way of introducing prospective officers to naval life, but, of course, history tends to record the astonishing successes. A notable example is Thomas Cochrane, whose uncle, Captain Sir Alexander Cochrane, entered his name on the books of several successive ships, starting when the boy was five. He didn't actually join his first ship until June 1793, a few months short of his eighteenth birthday, and went on to a distinguished career in the Navy. Britain's most famous Admiral, Horatio Nelson, also entered the Navy in breach of the regulations. He really did serve all his time at sea, but he joined his first ship (commanded by his maternal uncle, who had offered to take the lad off the family's hands) at the age of twelve, a year earlier than the regulations allowed since he was not the son of a naval officer. No less surprisingly, to modern eyes, is the familiar story of how Nelson's father, unable to accompany him further than London, put the twelve-year-old boy on the Chatham coach and left him to find his own way to the ship and make himself known. Nelson, of course, made a success of his new life, and was promoted to Lieutenant at the age of eighteen. Technically, this was another breach of regulations since there was supposed to be an age limit of twenty for this step. But the age limit was often ignored for young gentlemen of outstanding ability or those with powerful influences (what the officers of the time referred to as 'interest'). The most glaring example of this was the case of John Rodney, the son of Admiral Lord George Rodney, who joined his ship as a Midshipman at the age of fifteen (having served no time at sea) and was promoted to Lieutenant a few days later. A more entertaining example is the case of one Charles Adam, who 'passed' his examination for Lieutenant while serving in the East Indies, returned to England as an Acting Captain, was then informed that there had been an irregularity concerning his promotion, took the examination again and was promoted to Lieutenant for the second time, all before he was twenty.[15] It did him no harm, though; he ended up as First Sea Lord.

It was in order to bring some sort of regularity to this confusion that the Admiralty decided to set up a Royal Naval Academy in Portsmouth 'for the better education and training of up to forty young gentlemen [a year] for H. M. Service at sea'. The appropriate orders were issued in 1729, but the buildings to house the young gentlemen and their teachers had then to be erected, so the Academy did not open until 1733. The system of a letter of recommendation from the King was dropped in 1731, as the Academy was intended to replace that method of entry. It catered for only a minority of prospective Midshipmen joining the Navy (no more than 10 per cent); but it was always understood by the Admiralty that these were the candidates being groomed for higher command. This, of course, did nothing to make them more popular, either with their peers or their superiors, when they eventually joined their ships, after leaving the land-based establishment. But even with this experience behind them, they still had to serve at sea for a couple of years as volunteers before being rated Midshipman.

At this time, the age limits for entry to the Academy were between thirteen and sixteen years, and although the time to be served at sea before being allowed to qualify for promotion to Lieutenant was officially six years, the time spent at the Academy would count for two of those years. Since the maximum time a young gentleman could spend at the Academy was three years, this meant that in practice the earliest that a boy who entered at thirteen could be examined for promotion to Lieutenant would indeed be at the age of twenty. The Academy went through good years and lean (mostly lean) in the rest of the eighteenth century, with perhaps the most notable reform coming in 1773, from which time the sons of naval officers were admitted free of charge. But it was reconstituted in 1806 as the Royal Naval College. In that form, it lasted only until 1837; but among those who passed through the College during this brief incarnation was Robert FitzRoy, whose record marks him out as one of the best students ever to attend either the old Academy or the new College.

From 1806, there were seventy places available at the College each year, forty reserved for the sons of commissioned officers in the Navy, the rest for 'the sons of officers, noblemen and gentlemen'. From 1816 onwards, the College was reorganised along naval lines, rather than those of a school, with a Captain in charge of the establishment and two Lieutenants as his assistants, in overall charge of both students and schoolmasters, and it was under this regime that FitzRoy came on board in 1818. Although only twelve at the time, he was not under age, as the minimum age for entry to the College had been reduced since the inception of the original Academy. For what it is worth, this means that he joined the Navy at almost exactly the same age that Nelson did, although in distinctly happier circumstances. Lacking anything in the way of reminiscences from FitzRoy about his own time at the College, we can best get an insight into what life must have been like for him then, as we have mentioned, from the memoir of his near contemporary, James Sulivan.

Sulivan was born on 18 November 1810, at Tregew, near Falmouth, in the west of England. He was named after his maternal grandfather,[16] Admiral Bartholomew James, and his father was a naval Captain, who had three brothers also serving in the Navy. But the Captain had no income apart from his pay, and a large family. Some idea of how hard-pressed he was to support the family can be gleaned from the fact that he had to turn down the offer of a knighthood, because he would have been unable to pay the fees which the Crown then required from recipients of the honour. It was a combination of the family naval tradition and financial hardship that led young James to take up one of the free places at the College in 1823, when he was twelve years and three months old. The age for entry at that time was officially between twelve and a half and thirteen and a half; James was the eldest of four brothers who (like their father and uncles) all became naval officers. There was also another factor which made entry through the College potentially beneficial to a young gentleman with ability but nobody to look out for his interest. As Sulivan explains in his autobiographical note, 'there was also the chance of winning a medal at the college, which gave a prospect of promotion on passing for lieutenant'.

This was looking years ahead, but to the most important step in the career of any naval officer, the one from Midshipman to Lieutenant. After serving his time at sea, when he was considered ready a Midshipman would take an examination which, if he passed, would qualify him to be promoted to Lieutenant. But the crucial point is that passing the examination did not ensure the promotion, which would occur only if a vacancy became available. In the peacetime Navy, with fewer ships than during the Napoleonic Wars and no deaths from enemy action, such vacancies were rare – and when a vacancy did become available, a young gentleman with interest would be more likely to get it than a penniless 'passed' Midshipman, no matter how strong his family tradition of naval service. Men without interest, Sulivan tells us (and the records confirm):

> were sometimes ten, eleven and twelve years after passing before they were promoted, while they saw every one with interest, or as the saying was then in midshipmen's berths, 'with handles to their names,' promoted as soon as they had passed their examinations.

A son of Captain Pellew had won the second medal [that is, second in his year] at this college, and this was always held up to me as an example. My father had a large family; therefore it was a great object to him and to all poor naval officers to get so good an education from the navy free of cost.

But without that medal, even with the benefit of the College education Sulivan could end up as a 'passed' Lieutenant, still with the rank of Midshipman[17] at the age of twenty-five or even thirty, with little pay or status but well able to deal with the safe running of a ship, serving under officers who might be half a dozen years younger, with less ability and experience, who had been promoted early simply because they had 'interest'. Walker[18] quotes a letter from a Midshipman, written in the early nineteenth century, which underlines the point:

It not infrequently happens that a lieutenant is not nearly so old or so good a sailor as a mid in the same ship (which is the case here) & of course it is disagreeable to be ordered about like a dog by a man much younger than yourself who has been promoted not for any merit of his own but merely because he is Sir Somebody Something or because his interest lays in the present Ministry or something of that sort.

FitzRoy, of course, never had such worries – but, as we shall see, he had the ability and application which would have guaranteed his promotion even if both his grandfathers and two uncles had not been Lords.

Like Sulivan, FitzRoy would have had to pass a fairly straightforward examination before admission to the College. Sulivan describes his as involving only questions in arithmetic and geometry (Euclid), which he passed first out of an intake of twelve boys. The College education itself was far from being entirely devoted to practical matters and subjects such as navigation, and it is clear that the authorities wanted to turn out reasonably well-rounded gentlemen who would be a credit to the Navy no matter how high a rank they achieved and no matter what circles they might move in:

The head of the studies was the Reverend Professor James Inman, D.D., author of the work on navigation, under whom were three assistant-masters for mathematics: first, Peter Mason, M.A.; second, Charles Blackburn, M.A.; and third, Mr. Livesay. The preceptor, the Rev. W. Tate, M.A., took the classical classes, history, geography, and English. French was taught by M. Creuze, a French *émigré*. We were also taught fencing and dancing. The forenoons were given to mathematics, the afternoons to French and drawing, the latter taught by a very superior master, Mr. J. C. Schetky. There were also classes in naval architecture, which were taken by Mr. Fincham, the master-builder of the dockyard. We began geometry with Mr. Livesay; but no boy could get on unless he studied in his own cabin and at the dining-room tables in the evenings.[19]

The 'cabins' were in fact rooms in the College buildings – luxurious accommodation compared with what the young men would have to get used to when they went to sea. There seems to have been an element of bullying in the College, with some of the less studious students harassing those who wanted to work hard, at least in a mild way; but as no boy was permitted to enter the cabin of another boy, there was always somewhere safe to retreat to. A keen student certainly needed somewhere of his own to work. The workload described in the passage from Sulivan above was just for the new boys; at the other end of the scale, the official syllabus which laid down the subjects to be covered in the six half-years of the notional three years of the course concluded:

Fifth half-year – Fortifications, doctrine of projectiles, and its application to gunnery: principles of flexions [calculus] and application to the measurement of surfaces and solids: generation of various curves, resistance of moving bodies: mechanics, hydrostatics, naval history and nautical discoveries.

Sixth half-year – more difficult problems in astronomy, motions of heavenly bodies, tides, lunar irregularities: the *Principles* and other parts of Newton's philosophy, to those sufficiently advanced.

'Those sufficiently advanced' certainly included both FitzRoy and Sulivan, the latter benefiting from a friendly rivalry which brought him to some extent under the wing of the head of studies. Three months after Sulivan entered the college, Professor Inman's son, Richard, joined. As the two outstanding students of their year, encouraged by Professor Inman, they engaged in a competition which kept both of them hard at work. Although the course at the College was notionally three years long, students were encouraged to complete the syllabus and pass their examinations as soon as possible, with the understanding that out of each intake the one who completed the course in the shortest time was awarded the first medal, and the next received the second of the two medals awarded each year. Receiving the first medal ensured that a Midshipman would be promoted to Lieutenant immediately on passing; receiving the second medal made such promotion likely, but not certain. Sulivan and Inman finished neck and neck in 1824, having taken a little over a year and a half to complete the course (about the same length of time that FitzRoy had taken), and Professor Inman asked the Admiralty to award two 'first' medals, but they refused, because it would have created a precedent. Professor Inman was told that he must choose between the two boys, but that on this occasion (and without creating a precedent!) the recipient of the second medal would also be earmarked for instant promotion on passing for Lieutenant. Inman gave his own son the first medal, and Sulivan the second, which Sulivan himself says was 'perfectly fair', since he had benefited from the rivalry with Richard and Professor Inman's resulting personal encouragement of both boys.

At that time, senior Midshipmen took the examination for Lieutenant at the Royal Naval College, and as part of the learning process the senior class of collegians, as they were known, used to sit the same paper in the same room, but a day later. The last time Sulivan did this before leaving the College, Professor Inman told him that 'there was a [former] collegian passing yesterday who [had] won the first medal – his name is FitzRoy; and he did what has never been done before: in passing for a lieutenant he got full numbers [100 per cent correct marks], and I hope when you pass for lieutenant you will do the same'. Hardly surprisingly, on passing FitzRoy was immediately promoted to

Lieutenant. As it happens, Sulivan did eventually emulate FitzRoy's feat; but this was the first time their paths crossed directly, and the anecdote shows in what high regard FitzRoy was held from the beginning of his naval career, when he was already marked out as someone with much more going for him than a handle to his name. Almost immediately, Sulivan encountered FitzRoy again, when they were both appointed to the frigate HMS *Thetis*, where FitzRoy's influence on the young collegian was such that he later wrote 'to that I attribute much of my future success in the service'. At this stage, Sulivan was merely a 'College Volunteer', but at the time the term Midshipman was used widely to include such half-and-half creatures, and we shall follow this practice.

FitzRoy himself served at sea first in the *Owen Glendower*, which was posted to South American waters, joining the ship as a College Volunteer in the autumn of 1819, but becoming a Midshipman just a year later (remember that his time at the College officially counted towards his time at sea).[20] In November he wrote home from Madeira to tell his sister Fanny that 'I am very happy & comfortable on board & am sure I shall like sea life very well.'

In his published writings, FitzRoy has left us only one direct reference to this time in his life, when he is describing the approach of the *Beagle* to Rio in the second volume of the *Narrative*, and digresses to reminisce about his impressions on first seeing the same coast from the *Owen Glendower*:

High blue mountains were seen in the west, just after the sun had set, and with a fair wind we approached the land rapidly. The sea was quite smooth, but a freshening breeze upon our quarter carried us on, nearly thirteen knots an hour. Though dark as any cloudy tropical night, when neither moon nor star relieves the intense blackness – astern of us was a long and perfectly straight line of sparkling light, caused by the ship's rapid way through the water; and around the bows, as far forward as the bowsprit end, was dazzling foam, by whose light I read a page of common print. Sheet lightning played incessantly near the western horizon; and sometimes the whole surface of the sea seemed to be illuminated.

Here we have some of the essence of FitzRoy – the romantic, enjoying the beauty of the scene, coupled with the scientist, curious to find out if the phosphorescent light from the foaming sea is bright enough to read by. A couple of days later, he tells us, a party of Midshipmen

> were allowed to take a boat and enjoy a day's excursion in the beautiful harbour, or rather gulf. We landed on an island, which seemed to me like an immense hot-bed, so luxuriant and aromatic were the shrubs, and so exotic the appearance of every tree and flower. Years since elapsed have not in the least diminished my recollection of the novelty and charm of that first view of tropical vegetation.

He certainly sounds happy enough as a teenage Midshipman; returning in his narrative to the second *Beagle* voyage, older and perhaps beginning to feel the burden of command, he is less ebullient, and goes on to say that although

> the Sun shone brightly, and there were enough passing clouds to throw frequent shadows over the wooded heights and across vallies, where, at other times, the brightest tints of varied greens were conspicuous: yet I did not think the place [Rio] half so beautiful as formerly. The charm of novelty being gone, and having antici- pated too much, were perhaps the causes.

It may also have been, of course, that the beauty of Rio was enhanced by the contrast between the living conditions of a Midshipman and life in South America, and that on the later visit there was that much less contrast between FitzRoy's position as commanding officer and life ashore. The living accommodation shared by Volunteers and Midship- men proper, the Midshipmen's berth (strictly speaking, berths, since there would be several on a vessel of any size) was just as disgusting and uncomfortable as the picture painted by the novelists. It would be located just above the stinking bilges on the lowest deck of the ship, the orlop, at or below the waterline (depending on the size of the ship), with no source of natural light and only such air as filtered its way down from above. A space twelve feet square, with headroom of perhaps five and a half feet, would be considered generous accommodation for half a dozen young gentlemen. Walker quotes one Midshipman, writing home from a little ten-gun brig at the time of the Napoleonic Wars to tell his family that 'there will be 7 of us in a berth 8 feet by 6, therefore you may suppose we shall not have much room to spare'. Conditions would have been slightly better in the *Owen Glendower*, with a table in the middle of the room (which doubled as the surgeon's table in time of need, though happily for FitzRoy the wars were over by the time he went to sea) and each Midshipman allotted a space twenty-one inches wide in which to sling his hammock – seven inches more than the space allotted to ordinary sailors. The deck underfoot would have been home not just to the Midshipmen and the sea chests in which they kept all their possessions, but to the ship's rats and cockroaches, while food was of poor quality (to say the least) and never in sufficient quantity to satisfy the hunger of still-growing teenagers involved in a great deal of hard physical work. Although things improved slightly (only marginally) after the end of the Napoleonic Wars, an eight-year-old (!) "Volunteer" wrote in 1802 that:

> We live on [salt] beef which has been ten or eleven years in a cask, and on biscuit which makes your throat cold when eating it owing to the maggots, which are very cold when you eat them! Like calves-foot jelly or blomonge – being very fat indeed! We drink water the colour of the bark of a pear tree with plenty of little maggots and weevils in it, and wine, which is exactly like bullock's blood and sawdust mixed together.[21]

In 1832, things were scarcely any better, and Admiral Dalrymple Hay, serving off the West Coast of Africa, recorded that the salt beef provided for his ships had been salted back in 1809, twenty-three years earlier, and that even after it had been boiled it was so hard that it had to be grated with a nutmeg grater before it could be eaten. The ship's biscuit offered to them in one port for replenishing the ships was, he says, 'a caution', so swarming with maggots and weevils that:

In order to make it eatable – I will not say palatable – the bread bags filled with this biscuit were dragged out into the great square; on each bag was placed a fresh caught fish, the maggots came out of the bread into the fish, and the fish was then thrown away into the sea. A fresh fish then replaced the one thrown away, [and the process was repeated] until at last nothing more came out of the biscuit, when it was pronounced fit for food and served out to the squadron.[22]

Officers were able to purchase more palatable food, when they touched port, to improve their lot a little, but since Midshipmen were usually impoverished it was more likely that Lieutenants (and, of course, the Captain) would be able to take advantage of this. Inevitably, this led to the Midshipmen making raids on the supplies of their seniors, a practice which was not approved, and did not go unpunished if caught, but which seems to have been accepted by all concerned (remember that the Lieutenants had all been Midshipmen in their time) as part of the natural order of things, and to some extent condoned as encouraging initiative and enterprise among the young gentlemen.

All this, though, was merely the way of life of the Midshipmen when not involved in their nautical duties. Those duties did include some studying, subjects such as navigation and surveying, but were primarily practical, learning by experience in a highly hands-on way. The officer of the watch and other Lieutenants would run the ship, but with Midshipmen to assist them in ensuring the smooth running of the ship's routine. A good Captain, when there were no other ships about and the weather was fair, would even make a senior Midshipman take over the duties of officer of the watch, taking complete charge of the running of the ship – and then give a succession of orders for changes of course and so on (rather like a nautical driving test) to test the Midshipman's ability. When making or shortening sail, a Lieutenant would be in charge of each mast, with at least one Midshipman working aloft, and expected to set an example of speed and efficiency, even if he might be a raw teenager 'in charge' of a gang of burly seamen. When the order came to go aloft, it was expected that the young gentlemen would be first to their posts in the tops, whatever the conditions of wind and weather, playing a full part in making or taking in sail, reefing, or whatever. Sail drills were practised assiduously when in harbour or in calm conditions at sea, and ships of a squadron would compete with one another in races to see who could set or shorten sail fastest. It was a source of great pride, and a good career mark for the Captain, if the ship won these races consistently.

Perhaps the most important duty of a Midshipman, especially in terms of training for command, was to take charge of a boat. This might involve no more than ferrying men and stores from the quayside to his ship, riding at anchor in harbour, or an independent assignment, out of sight of the ship for weeks at a time in a strange part of the world on activities associated with surveying, or the suppression of slavery, gun-running and so on. In wartime, a Midshipman could even be given command of a prize, perhaps a captured merchant ship, with a handful of experienced sailors, a chart and a compass, and orders to make for the nearest friendly port. This certainly made the youngsters grow up quickly, with a delicate balance to be struck between being friendly with the men and being clearly seen to be in command. The first time a raw young middy was sent to the quayside in charge of a boat, it was almost certain that he would fall for the trick of allowing his men to step ashore to stretch their legs, only to find them comatose in a nearby tavern within a few minutes (the speed with which a British 'bluejacket' could drink himself insensible apparently had to be experienced to be believed). But as long as no drink was to hand, naval discipline and tradition mostly ensured that once the Midshipman had learned the ropes, and provided he didn't fall into the trap of trying to exert authority by bullying his men, things proceeded smoothly. Walker quotes from an officer serving in the 1860s:

On one occasion, the pinnace and the cutter were away for a fortnight. Lieutenant Miller in the pinnace, and I in the cutter, anchored under the lee of Cape Guardafui awaiting dhows rounding the cape. One morning I was ordered to round the point and land a man to look out from the cliff for slavers coming up from the south. During the afternoon it came on to blow quite suddenly, and in attempting to land to take off the look-out man, the wind, sea, and current drove us on the rocks. We were wrecked, and the boat smashed to pieces. We (14 in all) were stripped, ready for *sauve qui peut*. We dug ourselves into a small patch of sand for the night and the sailors treated me, 16 years of age, as a babe, and huddled round me and each other to keep warm. The next day, fortunately, was the rendezvous for the ship to meet us, and she came to the rescue. There was a fairly heavy sea running, making it impossible to land a boat, so a grass line with lifebuoy attached was taken on shore from a boat by a Krooboy. The latter escorted us through the surf, where we were hauled on board the boat, I being the last as officer-in-charge.

That neatly sums up the relationship between a popular Midshipman and his men – a combination of a young lad, learning the ropes, who they would look out for in a disaster, and a King's officer, in command, who automatically left the place of danger last.

A pinnace and a cutter are two kinds of small craft; one of the other favourite craft for such independent expeditions was a whale-boat, which got its name because it was, indeed, based on a design used by whalers. These were light, fast boats which could be rowed but also carried a mast and sails, with a centreboard that could be lowered to provide a keel which enabled them to beat upwind efficiently, or raised to allow the boat to run ashore in shallow water. As we shall see, such boats were particularly prized in surveying work, and one whaleboat in particular played a dramatic part in FitzRoy's life. We can be sure that all of these activities, from working in the tops during a storm, to taking over as officer of the watch and trying desperately to remember the right words of command to ensure the changes in course and sails demanded by the Captain, to command of a whaleboat on an independent expedition, would have come FitzRoy's way during his time as a Midshipman. But there were also opportunities for a well-connected young gentleman from an aristo-cratic background to see something of the world beyond the confines of his ship.

FitzRoy's letters home include mention of a spell of several weeks living ashore at Buenos Aires early in 1820, going out riding, hunting and so on. On 1 July that year, four days short of his fifteenth birthday, he writes to Fanny that he is learning Spanish, a language 'which, after French, is a very useful one to a sailor. I should wish to learn French first, but as I have the opportunity before me for the other must take advantage of it.' Here is another early hint of the hard-working FitzRoy, occupying his spare time usefully rather than in idle pursuits. By 28 November, there is more exciting news – 'I shall not get any more letters for a long time as we are going round Cape Horn.' It would be the first of many journeys around that notorious Cape, but as he later wrote to Fanny 'that "*Dangerous* and *troublesome*" (as some people style it) voyage round Cape Horn, which has been so much talked about, was performed . . . without any difficulty'. And then, in the same letter from Valparaiso in Chile, on 15 March 1821, we get a hint that young FitzRoy might, on occasion, have been a bit of a prig: 'I saw 5 Collegians in here, and *very little credit they did* the College, so little that I half *cut* them and I more than half affronted them by doing so. But that I care nothing for.'

In another letter, he describes a long journey inland, the beautiful countryside and the mountains, and a visit to Santiago. But the arrival of news from home is always a high point in his life: 'I wish you could see us when a packet arrives, every body so anxious to know whether there are any letters for them, & the long faces of those unfortunates who receive none, which, *many* thanks to my kind Correspondents has not yet been my case.'

FitzRoy spent two and a half years on board the *Owen Glendower* before her return to England, and a similar period in the *Hind*, seeing service in Mediterranean waters before returning to England on board the *Cambrian* (which he joined on 25 April 1824 and left on 5 July the same year) to take his examination for Lieutenant, which, as we have seen, he passed with full marks in August. His promotion to Lieutenant was dated 7 September 1824. Each time he changed ship, he had to take with him a certificate of competence from his Captain, couched in stereotyped, formal language; without these, he could not even sit the examination for Lieutenant. The one given to him on leaving the *Owen Glendower* reads:

These are to Certify the Principal Officers and Commissioners of His Majesty's Navy that Mr Robert Fitzroy has served as a Volunteer per order onboard His Majesty's Ship Owen Glendower under my command from the nineteenth day of October 1819 to the nineteenth day of October 1820 then as an Admiralty Midshipman to the third day of April 1822 during all which time he behaved with diligence attention and sobriety and was always obedient to command.

These were clearly happy and productive years. On 8 June 1822 he writes to Fanny from Gibraltar that 'I am giving all my spare time to Italian & a little French,' and from the Bay of Naples on 11 July that year, that 'altogether I am *very well off indeed*'. He delights in visits to Pompeii and Herculaneum, but gets little pleasure out of the balls held in Malta while the *Hind* is there: 'The fact is that in a garrison town like this there are an infinity of officers, & very few ladies who can dance, which few are generally engaged long before, and a Midshipman is a contemptible sort of fellow when a Colonel or a Captain is to be had.' Frustrating for a seventeen-year-old – but there was always work to keep him occupied. On 30 May 1823 he tells Fanny: 'I believe (*between you & me*) I get on better than *others*, but I always fancy that I might do more still, in the 5 or 6 hours leisure which is all I get of a day.'

Visits to the islands of the Eastern Mediterranean, including Cos and Rhodes, help to broaden his education, and back in Malta on 26 September he describes for Fanny a visit to Phasos: 'You have no idea of the beauty of part of the country which belongs to the Turks, & how little idea they have of enjoying it.' And from Smyrnia, where the ship had a long stay in harbour, he tells how he has been learning fencing and dancing from an Italian master – attributes that no doubt came in handy when he was promoted to Lieutenant. Back in England, FitzRoy kept up his correspondence with Fanny, who married George Rice-Trevor, known to his friends as 'Rice', the twenty-nine-year-old heir to the third Baron Dynevor, on 27 November 1824; they would have three children, all daughters, over the next twelve years. Something Robert did must have particularly pleased her, since in a letter from the *Thetis* in Devonport, simply dated 'Thursday', he says:

Dearest Fanny,

Your letter gives me great pleasure for I know of no satisfaction
equal to what one feels when conscious of having pleased a person,
tho' I suppose it is only '*gratified vanity*' after all.

Gratified vanity or not, this is one of our first insights into what would
be a driving force in FitzRoy's life – pleasing people, and doing good
for others. The same letter describes his accommodation on board the
ship:

[I] have made my Cabin *most* comfortable – quite a little Paradise
– I [have] been sadly expensive in the book line & have I flatter
myself a complete library in miniature (excepting one or two works
which I have ordered). You will hardly think that in a place 6½
feet square I stow – a broad Chest of Drawers 1 trunk – a large
table – washing stand 1 or 2 hats 2 cloaks, sticks & umbrellas
[illegible] & Guns, and upwards of *four hundred volumes*!

It was as well that he was so well provided for since he would soon be
back in South American waters, for a long time. South America was of
particular interest to the British in the decades following the Napoleonic
Wars, because those wars had led to the collapse of the old Spanish
and Portuguese empires in the region, and the emergence of newly
independent states. The United States, particularly under the presidency
of James Monroe, who served from 1817 to 1825, developed a policy
known as the 'Monroe Doctrine' (formally spelled out in 1823), which
asserted that any further European colonial ambitions in South America
would be seen as a threat to the United States, but that the United
States would not interfere in European political affairs if Europe kept
out of South America – kept out politically, that is. But there was a
difference between colonial intervention and trade, and Britain in
particular, the greatest seafaring power of the time, was determined to
open up trade with South America, whether the United States liked it
or not. So Britain kept a substantial naval presence in South American
waters in those decades, and one of the important duties of that naval
presence was to carry out surveying work, since such charts as the

Spanish had drawn up were regarded as state secrets, and never made public (and copies of those charts that had been acquired by the British anyway were notoriously inaccurate). As an official Admiralty historian later wrote, 'trade [is] the hidden dictator of surveying'.[23] All of which explains why both FitzRoy and Sulivan, serving together on board HMS *Thetis*, were soon to find themselves in Rio.

But Sulivan very nearly didn't serve on *Thetis* at all. The Captain of the ship, Sir John Phillimore, was one of the old school of seafarers who was suspicious of the new way of educating prospective officers, and had recently told Inman that he would refuse to accept any collegian posted to his ship. By chance, when Sulivan arrived on board Captain Phillimore was on leave, and the ship was in the charge of the First Lieutenant, Drew, and the Second Lieutenant, Cotesworth. Cotesworth was the son of another naval captain, who was a friend of Sulivan's grandfather; he warned Sulivan about the Captain's prejudice against collegians, but promised to do all he could to keep Sulivan on board. When the Captain returned, he summoned Sulivan for an interview in which the young man, still only fourteen years old and forgetting his lowly place in the naval pecking order, spoke up strongly in support of the College. The Captain's initial angry reaction was to order him out of the cabin; but it turned out that he had been impressed by the boy's spirit, and decided to take him on a trial basis. The trial was a success – after two months at sea, when the *Thetis* returned to Portsmouth Phillimore actually asked for two more collegians to make up his quota of young gentlemen.[24]

This willingness to adapt to changing circumstances was actually more typical of Phillimore than his early instinctive objection to collegians. In 1823, when he had commissioned the *Thetis*, Phillimore had obtained permission from the Admiralty to reduce the daily rum ration for the crew from half a pint per man to a quarter of a pint, in return for increasing the allowance of other luxuries. In spite of initial persecution of the crew by men from other ships, who referred to the men from the *Thetis* as 'tea-chests' and fought bitterly with them when ashore, the experiment proved a great success and was in due course extended throughout the Navy. Sulivan recalled that:

Later, when on detached service from the *Beagle*, and in boats for many weeks, while sitting round the fire at night, smoking and drinking quantities of tea, one of the oldest seamen in the ship, a petty officer, whom I always selected for my coxswain in those boat expeditions, used to tell the men that he looked back with shame and sorrow to the days when he helped others to attack the *Thetis*'s ship's company because they consented to try reducing the allowance of rum one-half, and he used to explain to the other men the great advantages of the change.

The more enlightened and forward-looking aspects of the way Phillimore ran his ship clearly influenced both Sulivan and the newly promoted Robert FitzRoy, now aged nineteen and with five years' genuine seagoing experience behind him, who joined the *Thetis* in the autumn of 1824 as its junior Lieutenant, a couple of weeks after Sulivan had joined and just before the cruise on which Phillimore changed his mind about the value of collegians. Seniority among officers of equal rank was (and is) decided by how long the officer has held that rank, so on the day of his promotion FitzRoy was the most junior Lieutenant in the Navy, not just on board *Thetis*; but early in 1825 an even more junior Lieutenant joined the ship, and FitzRoy began his inexorable rise up the chain of command.

Almost everything we know about the four years FitzRoy spent as a Lieutenant in *Thetis* comes from Sulivan's memoir. At that time, a frigate was always kept handy at Plymouth in case a dignitary such as an ambassador needed a ship to take him to his appointed post, or for other urgent duties. *Thetis* filled that role, for a time, and in between such duties cruised the Cornish coast looking for smugglers. Sulivan says that all of the Lieutenants were good officers, but that:

Lieutenant FitzRoy was one of the best officers in the service, as his subsequent career proved. He was one of the best practical seamen in the service, and possessed besides a fondness for every kind of observation useful in navigating a ship. He was very kind to me, offered me the use of his cabin and of his books. He advised me what to read, and encouraged me to turn to advantage what I had learned at college by taking every kind of observation that was useful in navigation.

The skills of seamanship and navigation required in beating up and down the rocky Cornish coast in gales coming in off the Atlantic in winter would also come into their own when both men were serving in the *Beagle* off Tierra del Fuego, and the young gentlemen of the *Thetis* learned the practical side of their profession as thoroughly as FitzRoy had when he was a Midshipman:

> Sir John Phillimore gave all the midshipmen a thorough practical training aloft. Every afternoon when the weather permitted it, the officer of the watch had to assemble the midshipmen an hour before the evening meal; and when we had taken in the lighter sails – flying royal and mizzen-topgallant studding-sail – if they were set, we had to take the first reef in the topsail, come down and hoist it; then the second reef, and come down and hoist it; then the third reef, doing the same; and then shake out the reefs singly. So we had to come on deck six times and hoist the topsails, and, if required, set the light sails above; and if the officer of the watch was satisfied with the way we had done it he sent us down to our tea.

And all this, remember, for young gentlemen who were typically between fourteen and eighteen years old. By the time they made Lieutenant, they were as at home in the rigging as they were on deck.

But there was more to being a King's Officer than seamanship and navigation, as a dramatic incident at the end of 1825 testified. Earlier that year, *Thetis* had been sent first to Lisbon and then into Mediterranean waters, before returning, homeward bound through the Straits of Gibraltar in mid-December. They found that a fierce gale had driven several large merchant vessels from their anchorage in the bay, driving them aground on the Spanish side. The Spanish had taken the opportunity to seize the ships, and unload part of their cargoes in lieu of customs duties, since they had (all too literally) landed in Spain. The Governor of Gibraltar, Sir George Don, refused to intervene for fear of causing a war with Spain. But Captain Phillimore was made of sterner stuff, and sent a large party in all the ship's boats to retake the merchantmen. Backed up by a few dozen mounted officers from the Gibraltar garrison, who (acting on their own initiative) made a show of force at the land border, this initiative was enough to force the Spanish to back down, without a shot being fired. The unloaded stores were reloaded, and the various vessels re-floated – no mean task, since 'the vessels were more than their own lengths from the water, and were all many feet in the sand. We had to dig channels to them, the men working up to their necks in water.' And this, remember, in midwinter; there were inevitable consequences. 'During the digging out of the vessels on the beach we sent several men back to the ship with various forms of pulmonary complaints, and one of the best first-class petty officers died of inflammation of the lungs.' But the mission was accomplished, in the best naval tradition. FitzRoy, as one of the Lieutenants of the *Thetis*, must have played a large part in all of this, although Sulivan does not go into details.

The ship finally reached Plymouth again in March 1826. Her next voyage was to take the newly appointed British Ambassador to the Argentine, Lord Ponsonby, and his retinue across the Atlantic. This meant building several temporary cabins on the *Thetis*, but on arrival in Rio it was decided to transfer the Ambassador's party to another ship, the *Doris*, which was heading south around Cape Horn on another mission, so that *Thetis* could return home to England. This meant a delay in Rio while the temporary cabins were dismantled and transferred to the *Doris*, and while this was going on a party of Midshipmen got permission from the Captain to take a boat to explore the islands of the river. At the outset of this expedition, almost as soon as they had got ashore one of the young men, George Wodehouse, had an accident with his shotgun, and opened up his arm from the palm of his hand to the elbow. It was touch and go whether he would survive, but the incident provided another insight into the way Captain Phillimore set an example to his junior officers, such as Sulivan and FitzRoy. He insisted on retaining what had been Lady Ponsonby's cabin for the wounded youth, and himself slept in a cot outside the cabin door. Although the other Midshipmen took it in turn to sit by Wodehouse, they were not allowed to touch his damaged arm; 'when our wounded messmate was suffering in the night we had to call the captain. I have known him sit an hour by his bedside, holding the arm in his two hands, trying to ease the pain.'

The *Thetis* sailed for home on 14 August, and Wodehouse survived. The ship itself was lucky to survive an encounter with a severe storm at the end of September, while still 400 miles from the Scilly Islands. Sulivan describes how boats were smashed, one man swept away and drowned, and how 'a sea struck her on the starboard side, and stove in two main-deck ports, and the water flooded the main-deck, pouring down every hatchway on the lower deck'. Even under these conditions, men (and boys) had to go aloft to adjust the sails, and the value of all their training became apparent. Lieutenant Cotesworth, overseeing this work from the main-top, on the tallest mast of the ship, could not make himself heard on deck because of the howling wind, so Sulivan was sent up to him to act as a messenger:

> As soon as I was in the top I was struck by the fact that when the ship was upright in the hollow of the sea the height of the sea hid the horizon. My eye was then sixty-four feet above the hollow. Directly I was sure of it I came down and told Lieutenant FitzRoy, who went into the main-top. His eye was sixty-five feet above the hollow. He states the fact in his account of the *Beagle's* voyage, and expresses his belief that it was the highest sea ever measured.

It is also a measure of the nonchalant skill of officers of the calibre of Sulivan and FitzRoy. It's one thing to work aloft in a tempest out of duty or the necessity of ensuring the survival of the ship; but these men had the time and curiosity to notice what was going on around them, and in FitzRoy's case to casually climb the rigging to see for himself, even though his duty did not require him to be aloft, making scientific observations of the size of the swell which might at any moment break against the ship and send it to the bottom.

A few days later, *Thetis* was anchored off Spithead, and on 28 October the ship was paid off. Captain Phillimore's opinion of the collegians was now so high that he asked Admiral Sir Edward Codrington, who was sailing in HMS *Asia* to take command of the Mediterranean fleet, to take all four with him; the Admiral agreed, but in spite of the offer two of the young men found what they regarded as an even better opportunity impossible to resist. On 29 October *Thetis* had been recommissioned, with Robert FitzRoy reappointed as one of the Lieutenants. 'Had it not been for Lieutenant FitzRoy's reappointment to the ship, we should all have accepted the offer, and should have been in the battle of Navarino. Baugh and Wodehouse accepted the offer, but Hamond and I preferred to remain in the *Thetis* with Lieutenant FitzRoy.' Good officers such as FitzRoy often attracted such loyalty, and this is yet another indication that FitzRoy was indeed an out-standing officer, even as a Lieutenant. In Sulivan's case, the loyalty would be amply rewarded before too long.

Under a new Captain, Arthur Batt Bingham, the ship was fitted out before returning to her duties in February 1827 – first taking a dignitary's party to Bermuda, and then going on to the River Plate to join the ships of the South American station under the command of Admiral Sir Robert Otway at Montevideo, where the Brazilians and the Argentinians were fighting for control of Uruguay, continuing a dispute left over from the rivalry of colonial days between their Portuguese and Spanish former rulers. The British kept a watching brief over the situation, ready to turn any opportunity to their advantage.

While serving in South American waters, far from home, FitzRoy learned that his father was seriously ill. The frustration of being far away and unable to help colours many of his letters. But it was also here, on the South American station on 11 December 1827, that FitzRoy and Sulivan first encountered the ship that would play a major part in both their lives. Sulivan's log for that day includes the note: 'Exchanged numbers with H.M. barque *Beagle*: 10.40 a.m., anchored near H.M.B. *Beagle*.'[25] But there is one last adventure concerning Lieutenant FitzRoy to recount before we take up the story of his time in command of the *Beagle*.

Because of the conflict between Argentina and Brazil, ships of both navies were always in need of men and, like the press gangs of England in the Napoleonic Wars, they were not above kidnapping sailors when they got the opportunity – a process known in those days as 'crimping'.

On 9 April 1828, six men from the *Thetis*, on duty ashore in Rio, were made drunk and carried off to serve on a Brazilian warship. By coincidence, just a week later Captain Bingham was returning in his boat from the harbour when he passed under the stern of the frigate *Imperatrix*, commanded for Brazil by a Danish mercenary, Captain Pritz. Bingham was hailed by the lookout on the *Imperatrix*, who turned out to be one of the crimped men; he informed the Captain that the others were also on board. Bingham immediately boarded the ship and, sending his boat back to the *Thetis* for reinforcements, confronted Captain Pritz, who had a hatred of the British dating back to Nelson's victory at Copenhagen,[26] with a demand for the return of his men. Hardly surprisingly, Pritz refused. But Bingham wasn't finished. The reinforcements from the *Thetis* were already being dispatched in two boats, one under Lieutenant Martin (who was senior to FitzRoy), and one under FitzRoy. Sulivan was in FitzRoy's boat; since it was the faster of the two, and the situation seemed urgent, Martin gave permission for them to go on ahead. As FitzRoy and his men appeared alongside the *Imperatrix*, Pritz backed down, and the crimped men were returned without violence.

Four months after this incident, without being involved in anything further out of the routine as far as we know, FitzRoy was transferred to the flagship, *Ganges*, to become Flag Lieutenant to Sir Robert Otway. Otway also had a Danish connection – he had been present at the Copenhagen battle, as Flag Captain (that is, Captain of the ship on which the Admiral flew his flag), and had tried to persuade the Admiral not to send the signal to which Nelson turned his blind eye. No doubt FitzRoy's part in the confusion of Captain Pritz met with Otway's approval, but this can hardly have been a factor in the appointment, which was a plum post for a young man just twenty-three years old, but exactly the kind of post, as an Admiral's right-hand man, that would be appropriate for someone with FitzRoy's aristocratic connections and 'interest'. FitzRoy knew why he had received the appointment, and told Fanny in a letter from Rio dated 29 August 1828, in which he also reveals that as far as he had been concerned all had not been plain sailing on the *Thetis*:

I am getting into good spirits again now that I know my Father is so much better – George gives me a very cheering account of him. What a year we have passed! . . . You will be very glad I know to hear that I am at this moment writing on board the Ganges – being Flag Lieutenant to Sir Robert Otway – and, I am *delighted* to say, have done with the Thetis – I hope for ever, – for if you knew the miserable hours I have passed on board her you would wonder I did not do much more foolish things than I did. I feel now a different person and beginning almost a new life – Now that I am clear of Captain Bingham, I tell you fairly that he is one of the emptiest *headed* vain men that ever annoyed his subordinates with fidgetty nonsense – but I believe him to be a very kind *hearted* man and good reason I have to say so, for his kindness to me in this Harbour went a long way towards saving my life when I was so ill with a sort of half Cholera Morbus half Dysentery. We have parted on excellent terms, and he has said many flattering things to the Admiral about me . . . I believe the next vacancy that occurs will be occupied by your brother Bob – what sort of a Captain will he make? Think you amongst the middlings – I will answer for his good *inclination* but not a word farther. Any one can make an every day sort of Captain but it is very difficult to go farther. The Admiral is very kind to me – indeed so is every body – particularly that good kind man – Lord Ponsonby – how glad I am he is come here as Minister . . . Fan how thankful I ought to be – as far as my interests here are concerned – here I am in a *Magnificent Ship* – holding a creditable comfortable place – having a large Cabin with light & air and on good terms with every one – pray tell my Father that my being here is in a great measure owing to Lord Ponsonby's kindness – for he & the Admiral are great friends – the Adm' had promised to take me into his Ship before but the Flag Lieut.^{cy} was Lord Ponsonby's doing.

In spite of this professional success, however, FitzRoy still longed in many ways to be at home, when news came that his father's health once again gave cause for concern. On 27 September he wrote that in spite of his joy in his new position 'still after all that has passed and is passing at home it is impossible to be happy'. But this seems to have been a fleeting bout of gloom, and almost immediately all thoughts of returning home were swept from his mind by another step up the professional ladder – his first command.

The fact that FitzRoy was an outstanding officer with proven ability and capacity for hard work, although by no means unimportant, was, as he acknowledges, hardly likely to have been the major factor in his appointment as Flag Lieutenant. Having both ability *and* interest, though, was a sure way to progress in the Navy of the time; and it soon turned out that FitzRoy had that other vital ingredient for success – luck. He was lucky enough to be in the right place at the right time when a vacancy tailor-made for a young officer of ability with interest appeared out of the blue. In October 1828 Philip Parker King, the Captain of HMS *Adventure* and in command of both *Adventure* and *Beagle* on a surveying mission in South American waters, sailed into Rio with the news that the Captain of the *Beagle*, Pringle Stokes, had shot himself in a fit of depression. Pringle asked Otway to appoint the *Beagle*'s First Lieutenant, William Skyring, as his replacement. But Otway chose to promote FitzRoy to the rank of Commander, and appoint him as Stokes's successor.[27]

While FitzRoy had been getting to grips with his role as a Flag Lieutenant, the *Thetis* had been away on various duties, and returned to Rio to discover 'that Lieutenant Robert FitzRoy had been promoted to commander, in the vacancy caused by the death of Commander Stokes, of H.M. sloop *Beagle*'.[28] The appointment had taken effect on 13 November. It was normal practice for a newly appointed Captain to take with him one or two junior colleagues, people he knew he could trust as he began to work with an unfamiliar set of officers, and in this case FitzRoy asked for Sulivan, who, with the Admiral's consent, was delighted to accept. 'I see by my log I joined the *Beagle* on December 15th, 1828, as a midshipman.' But Sulivan served on the ship in this capacity only until February 1829 when, in view of the need for him to return to England to take his examination for Lieutenant, he joined another ship for his passage home. As we have mentioned, he emulated FitzRoy's feat of achieving 100 per cent in the examination, in December 1829, and became a Lieutenant with seniority dating from 3

April 1830. A little over a year later, he would again be serving with FitzRoy on board the *Beagle*. But at last we can pick up the story of how and why that came about from FitzRoy himself, for just when Sulivan's account of his early life at sea ends, FitzRoy's *Narrative* of his time as Captain of the *Beagle* begins. The scene is set in a letter from FitzRoy to Fanny, from Rio, dated 23 November 1828:

> My Dearest Fanny,
>
> I know no one will be more glad, to hear of my promotion, than yourself – and you will be more pleased at hearing that it is given me by Sir Robert Otway – and not by the Admiralty. He has been very kind to me ever since I joined this ship – but this last act is indeed quite out of the common way – it is a *great gift*.
>
> A Death vacancy occurred in one of the two ships employed surveying the Southern coast of South America and Sir Robert has appointed me to fill it. What think you of your old Brother Bob being Captain of a 'Discovery Ship' – not perhaps quite that, but nearly so – for much of the coast is very little known. Now indeed I hope I shall have something worth writing about.
>
> It is not only promotion, but employment and that of the most desirable kind, for it opens a road to credit and character, and farther advancement in the Service. Providing I do not fail in my exertions.

He would not fail, and he would indeed have plenty worth writing about.

CHAPTER TWO

First Command

A S HIS ACTIONS were soon to prove, there is no doubt that, thanks to his upbringing and naval training, by the age of twenty-three FitzRoy was ready for command – even for what would soon be independent command, when the *Beagle* was dispatched on its own surveying work, away from the *Adventure*. But there were two aspects to command of a Royal Navy ship at that time. The Captain's job was to ensure that the orders he received were carried out to the best ability of the ship and the men under him, which required a great deal of initiative, with communications limited to seaborne dispatches received at long and irregular intervals, and the exercise of absolute authority to ensure that his own orders were obeyed without question by the officers and men under his command. Even a popular Captain – and FitzRoy was a popular Captain – was in some ways a remote figure, with almost god-like powers within the ship (including the power of life and death in the case of some crimes); he was the personal representative of the Sovereign, who in turn was the personal representative of God.

The other side of the coin was that in order to preserve this authority, the Captain could not be on intimate terms with anybody under his command, not even his most senior Lieutenant. They might be friends at one level, and the Captain might dine with his officers on occasion at their invitation, or they might be asked to dine with him. But there was nobody to confide in, especially about any worries or uncertainties. A Captain wasn't supposed to be worried or uncertain, and if he showed any sign of being so, maybe in a gale off Cape Horn, it would hardly be good for the morale and efficiency of the crew, upon which the safety of the ship depended. It was this loneliness of command, combined with the depressing weather, gloomy landscape and the difficulties of carrying out the work required of his ship, that had driven Captain Stokes to suicide. FitzRoy, with the example of his uncle, Castlereagh, also in his mind, was acutely aware of the danger of this aspect of the role he was taking on, and although he had no way to assuage any concerns he felt at the time of his initial appointment to the *Beagle*, they would play a significant part in the planning for his second voyage as her Captain.

One of the striking features of FitzRoy's first period in command of the *Beagle*, however, is the contrast between his attitude to the whole business and that of Stokes (incidentally, Stokes had served in the *Owen Glendower* in 1822 and 1823, so must have been known to FitzRoy). Stokes's journal,[1] although not exclusively gloomy, is full of references to bad weather, sickness among the crew, and the inhospitable nature of the landscape. For example: 'The effect of this wet and miserable weather, of which we had had so much since leaving Port Famine, was too manifest by the state of the sick list, on which were now many patients with catarrhal, pulmonary, and rheumatic complaints.'

And:

Nothing could be more dreary than the scene around us. The lofty, bleak, and barren heights that surround the inhospitable shores of this inlet, were covered, even low down their sides, with dense clouds, upon which the fierce squalls that assailed us beat, without causing any change: they seemed as immovable as the mountains where they rested ... as if to complete the dreariness and utter desolation of the scene, even birds seemed to shun its neighbourhood. The weather was that in which (as Thompson emphatically says) 'the soul of man dies in him.'

This was one of the last entries in his journal, written in June 1828, in the depth of southern winter, and clearly showing his state of mind. Captain King writes charitably that: 'Those who have been exposed to one of such trials as his, upon an unknown lee shore, during the worst description of weather, will understand and appreciate some of those feelings which wrought too powerfully upon his excitable mind.' But FitzRoy, exposed to exactly the same trials, and in command of the same ship with essentially the same crew, seems to have revelled in the experience, enjoying if not quite every minute of it then certainly by far the bulk of his voyaging around the southern tip of South America. His own journal records a triumph of optimism over adversity:

At dusk we pulled into a small creek, and secured the boats, hauling up the whale-boat on the sand. When too late to remove, we found the place of our bivouac so wet and swampy, that nearly two hours were occupied in trying to light a fire. Supper and merry songs were succeeded by heavy rain, which continued throughout that night and the next day without intermission.

It is hard to imagine Stokes leading his party in 'merry songs' under such circumstances! And try this, from 24 May 1829, again close to southern midwinter:

The last few nights have been so clear, that two or three of the men, and myself, have slept in the open air without any covering other than our blanket-bags,[2] and clothes. My cloak has been frozen hard over me every morning; yet I never slept more soundly, nor was in better health.

As well as accurately indicating which of the two Captains it would be happier to serve under, both these passages give the lie to the common perception that FitzRoy's journal is dull and not worth reading. It's true that his contribution to volumes I and II of the *Narrative* lack the literary flair of Darwin's Volume III, but Darwin was an outstanding writer, and to say that FitzRoy did not write as well as Darwin is by no means to suggest that he wrote badly. Volume II, in particular, is well worth reading even today, and it is surprising that it was out of print for more than a century until the edition edited by David Stanbury appeared in 1977.

Another small indication highlights FitzRoy's willingness to try out new things, his confidence in command from the outset, and his forward-looking approach. Under him on this voyage, the *Beagle* was the first Royal Navy ship to replace the word 'larboard' with 'port' (already in use in the Merchant Navy) for use in both written and spoken commands. This removed the possibility of confusion between 'starboard' and 'larboard', and now seems a small but obvious step. Nevertheless, someone had to be the first to take it, and that someone was the twenty-three-year-old Robert FitzRoy. By the time he had finished with the *Beagle*, he had also brought into the language of the Royal Navy another term, previously used by the merchantmen of the East India Company, and originally of Indian origin – the term 'dinghy' to refer to a small boat (previously described as a jolly boat).

By the time he took command, the *Beagle* and her crew had been well tried in South American waters. After carrying out the long, independent survey during which Stokes had become so depressed, at the end of July 1828 *Beagle* had rejoined *Adventure* in the Strait of Magellan at the unglamorously named Port Famine, a bay in the heart of the Magellan Strait which was the site of a failed sixteenth-century Spanish colony (hence the name) about which King wrote: 'This was the first, and perhaps will be the last, attempt made to occupy a country, offering no encouragement for a human being; a region, where the soil is swampy, cold, and unfit for cultivation, and whose climate is thoroughly cheerless.'

In these surroundings, amid the delight of renewing acquaintance with old colleagues and discovering that each ship had survived the separation intact, it seemed to King that Stokes 'was evidently much excited, and suspicions arose in my mind that all was not quite right with him'. He asked the two ships' surgeons to assess Stokes's state of health, and it was while they were on board *Adventure* preparing what turned out to be an unfavourable report that, on 1 August, 'a boat came from the Beagle, with the dreadful intelligence that Captain Stokes, in a momentary fit of despondency, had shot himself'. Stokes was still alive, even though the shot had been to his head, and the surgeons rushed to his aid. For four days he lay in a delirium in which 'his mind wandered to many of the circumstances, and the hair-breadth escapes, of the Beagle's cruize'. There followed a brief recovery, but in spite of the best efforts of both surgeons 'he then became gradually worse, and after lingering in most intense pain, expired on the morning of the 12th'.

As we have noted, Captain King gave the temporary command of the *Beagle* to Lieutenant Skyring, and he served efficiently in that capacity until October, when the *Adventure* met up with Sir Robert Otway, in his flagship *Ganges*, in Rio. FitzRoy was appointed to the command of the *Beagle*, with Skyring as his First Lieutenant. This was clearly not a decision that King welcomed, and he wrote:

> The conduct of Lieutenant Skyring, throughout the whole of his service in the Beagle, – especially during the survey of the Gulf of Peñas, and the melancholy illness of his captain, – deserved the highest praise and consideration; but he was obliged to return to his former station as assistant surveyor: and, to his honour be it said, with an equanimity and goodwill, which showed his thorough zeal for the service.
>
> Captain FitzRoy was considered qualified to command the Beagle: and although I could not but feel much for the bitterness of Lieutenant Skyring's disappointment, I had no other cause for dissatisfaction.

Clearly, though, King did not believe that, as of October 1828, FitzRoy was as well suited to the command of the *Beagle* as Skyring, who had already experienced the sharp end of surveying the southern tip of

South America in winter, and proved himself in command. And it must have been galling for Skyring to have the vital step from being subordinate to being in command taken from him after a few months.

Philip Parker King was not a man to suffer fools gladly. Then in his mid-thirties, he had already carried out survey work along the Australian coast (his father, incidentally, had been the first Governor of New South Wales), to such good effect that he had been elected a Fellow of the Royal Society, as well as being offered the South American survey. But if he had no time for fools, he was also quick to acknowledge real ability in others. It says much for the character of both FitzRoy and Skyring that the new arrangement worked not only smoothly but amicably, doubtless helped by the fact that FitzRoy soon proved himself to be a first-class seaman and surveyor in his own right, removing any doubts King had had about his 'silver spoon' treatment.

In one respect, FitzRoy was lucky; the *Beagle* was in need of a complete overhaul before being ready for the rigours of another surveying voyage, so for the first weeks of his command the ship was hove down and repaired, giving him a chance to get to know both ship and crew thoroughly in safe harbour, before experiencing any storms or other tests at sea. Any captain loves his first command, but there must have been some mixed feelings in FitzRoy's mind about the suitability of the *Beagle* for the tasks ahead. Although he describes her as 'a well-built little vessel, of 235 tons, rigged as a barque, and carrying six guns', he must have known that the particular design of ships to which *Beagle* belonged was sometimes referred to as coffin ships, because of the ease with which they could turn turtle if mishandled under adverse weather conditions, or as half-tide rocks, because they lay so low in the water that even moderate waves would sweep across their decks. As Sulivan put it, these ships required 'careful handling and management of sail', a potentially alarming prospect for a first command.

To modern eyes (though surely not to FitzRoy), the physical dimensions of the *Beagle* are no less alarming. She was just 90 feet long, with a maximum breadth of 24 feet 6 inches, drawing just 7 feet 7 inches forward and 9 feet 5 inches aft. Try pacing out an oblong 90 feet by 25 feet on a convenient piece of open ground, and consider whether you would fancy your chances of rounding Cape Horn under sail, in a

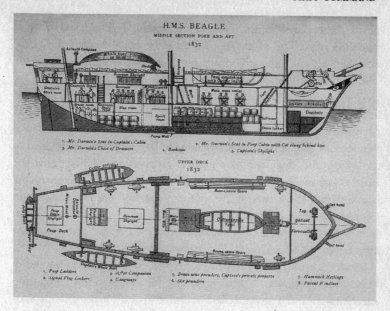

HMS *Beagle*, as refitted for the second voyage to South America.

gale in a vessel that size! To man the ship, FitzRoy had under his command, as well as William Skyring, just one other watchkeeping officer, a second Lieutenant (Lieutenant Kemp), as well as a sailing master, eight other officers (including, after Sulivan's departure, one Midshipman, John Lort Stokes, who was no relation to Pringle Stokes), a Sergeant and nine Marines, and 'about forty Seamen and Boys'. A grand total of some sixty-two people crowded into the tiny vessel, along with all their stores, food and equipment.

The rigging of the ship was just as important. When she was launched, at Woolwich, on the Thames, on 11 May 1820, *Beagle* was rigged as a brig, with two masts each carrying square-rigged sails (the name brig is a shortened form of the older brigantine). But the ship did not go into service at that time, and was laid up 'in ordinary' until the Navy had need of her. In 1825, when that need arose, she had been re-rigged as a barque, with a third mast (the mizzen) added at the stern end of the vessel, carrying fore-and-aft sails (the name is a shortened form of the older barquentine), but still square-rigged on the foremast and mainmast. The advantage of the barque rig is that the additional fore-and-aft sail, although only on one mast, provides more manoeuvrability, especially when beating into the wind.

The ability of both the ship and its new Captain to cope with extreme weather conditions was soon tested. On 27 December 1828, *Adventure* sailed south, leaving *Beagle* to complete her repairs and rendezvous with *Adventure* in Maldonado Bay, where King intended to go ashore and carry out astronomical observations to determine the precise latitude and longitude, part of a chain of such measurements made during the survey of the southern coast of South America. On 30 January, while safely at anchor in the bay, the *Adventure* was hit by a severe storm (known locally as a 'pampero'), which had been preceded by a fall in the barometer to 29.5 inches of mercury.[3] The squall which marked the onset of the pampero, accompanied by a sudden shift in the wind from northwest to southwest, laid the anchored ship[4] broadside on the water, and destroyed two boats on the shore; locals told Captain King that the violence of the wind, lasting for twenty minutes, was the worst for at least twenty years. And *Beagle*, King knew, should by now be on her way to join them.

On 1 February, *Beagle* arrived in the bay, having encountered the pampero at sea. The ship had lost both topmasts and many other spars, and had also lain on her beam ends in the wind, at risk of turning turtle and sinking. FitzRoy had saved her by lowering both anchors so that they acted as a drag, which brought the head of the ship into wind and allowed her to recover, although two men were blown from the rigging and drowned. In his *Weather Book* FitzRoy emphasises his failure to take sufficient precautions when he noticed the fall in pressure that preceded the storm:

In 1829, the writer was approaching Maldonado harbour (in the Beagle) when a *Pampero* was threatening. Signs in the sky, barometric evidence, and temperatures shewed what was coming, but want of practical faith in such indications, and impatience as a young commander, in sight of his admiral's flagship,[5] induced disregard, and *too late* an attempt to shorten sail sufficiently.

Topmasts and jib-boom were blown away, the vessel was just saved from foundering (being almost on her beam ends) by cutting away both anchors, and letting the cables run out to the clinch – which brought her head to wind and righted, – while two fine fellows, blown from aloft, swam hard for their lives, but were immediately overwhelmed by the sea – that was torn along, not in 'spoondrift' or spray, but in a dense cloud of broken water.

Sulivan, who had crawled from his cot, where he had been laid low with dysentery, describes how the ship was so far over on her side that he stood on the nearly horizontal mizzenmast, and saw 'the commander standing on one of the uprights of the poop-rail, and holding on by another upright', like a man clinging to a vertical fence, or the wall-bars of a gym. In the *Narrative*, FitzRoy describes his mixed feelings at saving the ship:

As the depth of water was small, and the ground tenacious clay, both anchors held firmly, and our utmost exertions were immediately directed towards clearing the wreck [i.e. wreckage], and saving the remains of our broken spars and tattered sails. Had we suffered in no other way, I should have felt joy at having escaped so well, instead of the deep regret occasioned by the loss of two seamen, whose lives, it seemed, might have been spared to this day had I anchored and struck topmasts, instead of keeping under sail in hopes of entering Maldonado before the pampero began.

This was clearly one of the key events that opened FitzRoy's eyes to the importance of understanding how the weather worked, and learning how to predict its behaviour. But although FitzRoy may have regarded his inability to read the signs accurately and act upon them at that time as a failure, it was, in fact, an impressive achievement to bring one of the 'coffin ships' through such a storm and safely into harbour. There is no hint in King's narrative or in the official records of any suggestion that FitzRoy was at fault; indeed, it seems likely that his seamanship in these difficult circumstances did much to allay any remaining doubts King might have had about his appointment.

FitzRoy himself was left with no illusions about the hazards of taking such a small ship on the expedition ahead of them. Although he makes only an oblique reference to the pampero, in a hastily written letter dated 16 February 1829 he writes what he clearly anticipates might be his last letter home:

My Dearest Fanny

I only write a few lines to wish you good bye and, to pray for a blessing upon you and your husband and children – Tomorrow morning I sail in the Beagle – in company with the Adventure – for the Straits of Magellan – At their entrance the Adventure will probably leave us – to work our way through and along the coast to the North & West as far as Chiloe ... for the last two months I have hardly had time to sleep ... We have been so hurried in fitting out and have had so much to contend with in weather that hustle & bustle have been the order of the day constantly...

After sending his best wishes to many friends and family, he concludes: 'God Bless you Dearest Fanny – Believe me Ever your most Affectionate Brother – Robert FitzRoy.'

King's little squadron was completed by the *Adelaide*, a schooner which he had purchased (after long and tedious correspondence with the Admiralty in London to obtain permission) in December 1827. Apart from the obvious advantages of having three ships instead of two to carry out the surveying work, a schooner was particularly suited to the task. With two masts, her foremast shorter than her mainmast, carrying large, flat-topped gaff sails, and a triangular sail extending forward from the foremast, all fore-and-aft rigged,[6] the schooner was highly manoeuvrable, could beat very effectively into wind, and required only a relatively small crew, made up on this occasion from the officers and men of the *Adventure* and *Beagle*. A Lieutenant Thomas Graves, from *Adventure*, was initially given command of the *Adelaide* by King. After they had carried out repairs to the damage caused by the pampero and another severe storm which struck them on 2 February, the three ships headed south via Montevideo, but were hit by further gales which washed away several boats (by the middle of March, King is noting ruefully that the expedition had lost eleven boats since leaving England, a severe problem for the inshore surveying work), and became separated. Although *Beagle* and *Adventure* soon found each other, they did not meet up with *Adelaide* again until the end of March, at a prearranged rendezvous.

After taking the opportunity to stock both of the smaller vessels with enough provisions to last them through the winter, on 1 April 1829, while *Adventure* set off to carry out the next part of her own survey, *Beagle* and *Adelaide* were dispatched to travel through the Magellan Strait,[7] surveying as they went, and up the eastern coast of South America to the island of Chiloé, at latitude 43° south (for comparison, Cape Horn is at 56° south and the estuary of the River Plate is at 35° south; five degrees of latitude corresponds to about 400 miles of straight-line sailing, much more, of course, if following the irregular coastline around the tip of South America).

Six months earlier, FitzRoy had been a Flag Lieutenant, a post that any of his contemporaries would have regarded as a cushy number, with the comfort and security of life on board the flagship in the warmer regions of the South Atlantic. Now, still not twenty-four years old, he was not only in command of the *Beagle* but also, as the senior officer, responsible for the *Adelaide*; two tiny ships heading into the depths of winter around some of the stormiest and most dangerous coastline in the world, going literally into unknown dangers, since the whole point of the expedition was to chart those dangers for the benefit of others. He revelled in it.

The two ships entered the maze of channels making up the Straits of Magellan through the Narrows at their eastern end, with only minor excitement – *Beagle*'s anchor cable snapped while they were waiting out the night, and FitzRoy had no choice but to shoot through the Narrows in the dark, on a strong tide. Inside the Straits, the two ships found plentiful supplies of cranberries and wild celery, an adequate supply of water, and were able to obtain guanaco meat (the guanaco is a relative of the llama) from the natives. This was FitzRoy's first encounter with Fuegians:

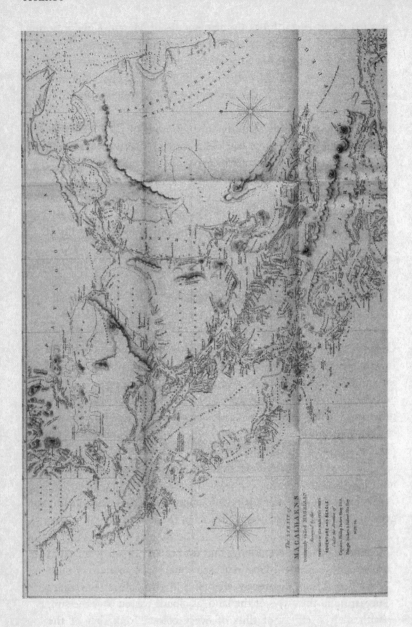

In the canoe were an old woman, her daughter, and a child . . . Their figures reminded me of drawings of the Esquimaux, being rather below the middle size, wrapped in rough skins, with their hair hanging down on all sides, like old thatch, and their skins of a reddish brown colour, smeared over with oil, and very dirty . . . Their canoes, twenty-two feet long, and about three wide, were curiously made of the branches of trees, covered with pieces of beech-tree bark, sewed together with the intestines of seals. A fire was burning in the middle, upon some earth, and all their property, consisting of a few skins and bone-headed lances, was stowed at the ends.

The young woman would not have been ill-looking, had she been well scrubbed, and all the yellow clay with which she was bedaubed, washed away. I think they use the clayey mixture for warmth rather than for show, as it stops the pores of the skin, preventing evaporation and keeping out the cold air. Their only clothing was a skin, thrown loosely about them; and their hair was much like a horse's mane, that has never been combed.

On 14 April the two ships anchored in Port Famine, and on the 16th Lieutenant Skyring, a more experienced surveyor than Graves, was sent on board the *Adelaide*; as Graves's senior, this meant that he was now in command of the ship with Graves as his First Lieutenant. The reason for the change was that *Adelaide* was about to depart to survey the Magdalen and Barbara channels, arms of the Magellan Strait already partially known from earlier surveys, while *Beagle* would explore other, lesser known channels between the islands, before the two ships joined forces once again.

On 19 April, setting out into the unknown, FitzRoy was clearly in good spirits:

I cannot help here remarking, that the scenery this day appeared to me magnificent. Many ranges of mountains, besides Mount Sarmiento, were distinctly visible, and the continual change occurring in the views of the land, as clouds passed over the sun, with such a variety of tints of every colour, from that of the dazzling snow to the deep darkness of the still water, made me wish earnestly to be enabled to give an idea of it upon paper; but a necessary look-out for the vessel, not having a commissioned officer with me who had been in the Strait before, kept my attention too much occupied to allow me to make more than a few hasty outlines . . .

The night was one of the most beautiful I have ever seen; nearly calm, the sky clear of clouds, excepting a few large white masses, which at times passed over the bright full moon: whose light striking upon the snow-covered summits of the mountains by which we were surrounded, contrasted strongly with their dark gloomy bases, and gave an effect to the scene which I shall never forget.

Anyone who thinks FitzRoy couldn't write has surely never read the *Narrative*!

Such weather never lasts long around Tierra del Fuego, and the more common meteorological observations in the *Narrative* refer to wind and rain, just the conditions to make even the tiny *Beagle* seem like a cosy refuge from the elements. Tierra del Fuego extends down to 55°S, roughly the same distance south of the equator that Labrador, in the Canadian province of Newfoundland, is to the north of the equator, which gives some idea of the conditions they were likely to experience as the southern winter deepened. But *Beagle* herself was too big for the detailed surveying work that was now required, which had to be carried out from small boats (including the invaluable whaleboat), in the teeth of whatever the elements could throw at them, including wind, snow and hail. Soundings had to be taken to ascertain the depth of the water, surveying instruments had to be set up on land to determine the positions of geographical features relative to one another, and astronomical observations had to be made (whenever the weather permitted) to locate the positions of those geographical features on the globe. Sometimes, the seamen would have to lug the surveying equipment up mountains in order to make observations from the peaks. On one occasion, FitzRoy tells us, a height 'ascended by Lieutenant Skyring was so steep, that the men were obliged to pass the instruments from one to another, at a great risk of their own lives; and when they reached the summit, the wind was so strong, that a heavy theodolite and stand, firmly placed, was blown over'.

After a couple of weeks of this kind of work in the area around where *Beagle* was anchored, FitzRoy prepared for a bigger expedition. Taking the whaleboat himself, with the cutter in the charge of Midshipman Stokes and a month's supplies, he left the *Beagle* in the hands of Lieutenant Kempe and set out on 7 May to explore an indentation on the northern shore of the western half of the Strait. It was in writing about this expedition that he waxed lyrical about sleeping out under the stars in midwinter; warming (if that is the right word) to the theme on 14 May, FitzRoy writes, 'so mild was the weather, that I bathed this morning, and did not find the water colder than I have felt it in autumn on the English coast; its temperature, at a foot below the surface, averaged 42°; that of the air was 39°'. More prosaically, they found that the indentation led to a huge, almost landlocked bay, which they named Otway Water; this then connected through a narrow channel (later named the FitzRoy Passage, by King) to another large bay, dubbed Skyring Water; but there was no connection back into the Magellan Strait and the Pacific Ocean. The surveying went well, and there were several cautious but friendly encounters with natives from different tribes, who were eager to barter guanaco meat and seal skins for knives and tobacco. On the return through Otway Water, however, the two boats survived an incident which again demonstrated FitzRoy's ability as a seaman and commander, and which gave him great satisfaction as an indication of the quality of the men under his command.

On 20 May 1829, the boats were heading east under what started out as 'a fine breeze' but soon developed into a gale, blowing them towards a flat shore with high surf pounding upon it. In the overloaded whaleboat, they had to lower all sail and point the boat with its head to wind and waves, rowing desperately to keep her off the lee shore; the cutter, which could still just carry a closely reefed sail, was sent off to beat to windward for as long as possible. The short winter day was already turning dark, and: 'At three o'clock, we were embayed, and about a mile from the shore. My boat was deeply laden, and as our clothes and bags got soaked, pulled more heavily. We threw a bag of fuel overboard, but kept everything else to the last.'

Although they caught a glimpse of the cutter, about three miles out, night soon fell, and 'having hung on our oars for five hours', FitzRoy began to think of beaching the boat to save the men, even if everything else were lost:

> It was not likely she would live much longer. At any time in the afternoon, momentary neglect, allowing a wave to take her improperly, would have swamped us; and after dark it was worse. Shortly after bearing up [to head for the shore], a heavy sea broke over my back, and half filled the boat: we were baling away, expecting its successor, and had little thoughts of the boat living, when – quite suddenly – the sea fell, and soon after the wind became moderate. So extraordinary was the change, that the men, by one impulse, lay on their oars, and looked about to see what had happened.

About an hour after midnight, ashore at last, the exhausted men were at last able to collapse into a deep sleep around a fire. 'No men could have behaved better than that boat's crew: not a word was uttered by one of them; nor did any one flag at any time, although they acknowledged, after landing, that they never expected to see the shore again.' FitzRoy's satisfaction was complete when, just before dawn, he was roused by Stokes, who had just brought the cutter safely in to shore.

There were no further alarms during the rest of the trip. On 1 June, the cutter was dispatched to rejoin the *Beagle*, with the whaleboat following more slowly, completing the survey. After a month of this independent existence, sleeping out in all weathers, suffering a near capsize, hauling surveying equipment up and down mountains, and so on, the crew of the whaleboat must have presented the appearance of a band of desperadoes;[8] but they still took pride in their appearance. On 8 June, 'as it rained heavily, we remained under such shelter as we could obtain; and prepared for our return to the Beagle, by making use of the only razor we had'. When the rain ceased, the clean-shaven crew sailed to the rendezvous with *Beagle*, having been away for exactly a month and a day. The following day, they were joined by the schooner *Adelaide*, with Lieutenants Skyring and Graves reporting that their work had been carried out without untoward incident and that all the men were well. FitzRoy's orders were now to take both ships to rendezvous with Captain King and the *Adventure* at the port of San Carlos on the island of Chiloé, off the west coast of southern South America;[9] but acting with the supreme confidence that he was often to show in the future, he decided it would be better to make use of the opportunity for *Adelaide* to make a close survey of the part of the Strait where he hoped there might be an opening to the Skyring Water.

FitzRoy was well aware of the possible consequences of this decision. Although his lieutenants agreed that this was a worthwhile expedition, if anything went wrong then as commander he would take the full blame. King was expecting both ships to sail together to Chiloé, and in particular for Skyring, his surveying work completed, to be back on board the *Beagle*, with Graves in command of the *Adelaide*. But: 'Much to the credit of Lieutenant Graves, he removed one weight, by volunteering to go any where I thought proper to direct, whether alone or with Lieutenant Skyring, and the necessary orders were forthwith given.'

This incident tells us a great deal about FitzRoy. His willingness to act independently, even if not exactly in accordance with the letter of his orders, if he thought that his overall mission required it, was in the tradition of the great British naval commanders, including Nelson. There were captains who, whatever the circumstances, would have stuck to the letter of their orders even if they saw there was an opportunity to do more, knowing that by obeying orders they would never be in the wrong. But those were not the kind of officers who became great leaders such as Nelson. It is not too fanciful to suggest that, especially given his undoubted popularity as a commander with the men under his command, FitzRoy would, had he been a captain in the Napoleonic Wars, have been one in the Nelson mould, a leader rather than a follower. That popularity, born out of confidence in the commander's ability, as well as from his personal charm, clearly influenced the behaviour of the men in the whaleboat during the gale in Otway Water, and equally clearly shows up in Graves's decision to 'volunteer' for the additional surveying mission. Of course, Graves had no choice but to go where FitzRoy told him; to do anything else would be a court-martial offence. But he could have done so reluctantly, registering his protest formally in the ship's log. The fact that he openly supported his young Commander's decision, even at the cost of once again being only the First Lieutenant, not the Captain, of the *Adelaide*, reinforces our image of FitzRoy as an inspiring team leader who had the confidence and respect of his men. So the two ships separated once again. With more surveying to be done en route, it was 9 July before the *Beagle* reached San Carlos, where the town reminded FitzRoy of a Cornish village. Just one untoward incident, a harbinger of things to come, occurred on the voyage – the precious whaleboat, hanging in its davits, was carried away by a sea breaking over the ship (the 'half-tide rock'), the result of 'a moment's neglect of the steerage'. But at least they would be able to get a new whaleboat made at San Carlos.

In San Carlos, *Beagle* was soon joined by the *Adventure*, but *Adelaide* did not turn up until 20 September, relieving mounting anxiety about her wellbeing. She had been delayed only by the thoroughness with which she had been completing her survey, and Captain King was pleased with the work carried out by the *Adelaide*, although this failed to find the sought-for link between Skyring Water and the Strait (whether he would have been pleased if the ship had failed to return safely is another matter, of course, and those weeks in San Carlos must have been anxious ones indeed for FitzRoy). With both the smaller vessels, in particular, in need of repairs and re-equipping after their arduous duties, there was a lengthy stay in the harbour, with plenty of opportunity for the officers to relax in the little town. During this stay, King received a significant and welcome dispatch. His original orders had been to proceed back to England westward, via New South Wales, where Britain had an infant colony, and the Cape of Good Hope; but he had requested permission to return by the direct route back around Cape Horn, and the dispatch brought news that his request had been granted. King decided to sail via Valparaiso and Port Famine to Rio de Janeiro, with *Adelaide* carrying out some more independent surveying work before joining *Adventure* at Port Famine, and *Beagle* proceeding independently on a much larger survey of the southern coasts of Tierra del Fuego (which is shaped like a triangle), before meeting the other two ships at the rendezvous in Rio by 20 June 1830. So it was that the *Beagle* set out from San Carlos on 19 November 1829, heading south and east on the voyage that would bring FitzRoy into closer contact with the Fuegians than he could ever have imagined, and which, as a result, would soon bring the *Beagle* back to these waters carrying a new passenger, the naturalist Charles Darwin.

The surveying work proceeded much as before, handicapped (as before) by the wind and the rain. On 17 December, FitzRoy noted that 'though the middle of summer, the weather was not much warmer than in winter', and the next day recorded 'a continuance of bad weather: no work was done in the boats this day. In the afternoon I tried to go up the mountain I had ascended on Tuesday, to bring down a theodolite which I had left at the top; but the wind obliged me to return unsuccessful.' When the weather did permit, FitzRoy used to find a safe anchorage for the *Beagle* along a particular stretch of coast, then send out one boat in each direction along the coast, taking one himself and putting one in the charge of the ship's Master, to do the detailed work. But one factor became increasingly different from the earlier voyage; the natives got more and more hostile as the *Beagle* progressed in her work, going beyond the petty thieving to which the ship's crew were accustomed.

On 21 December, with the ship at anchor in safe harbour at the northern end of an island in a bay, FitzRoy sent the whaleboat, under the command of the Master (Murray) and a Mate (Wilson, borrowed from the *Adventure*; one of those unfortunate souls who had passed for Lieutenant but not yet been given the substantive rank), to take measurements along the eastern part of the island. The party had provisions for four days, a tent and guns; but as the wind shifted and rose, bringing lowering dark clouds and heavy rain for Christmas, FitzRoy knew that it would be impossible for the boat to return, just as it was impossible to send further supplies to them by sea. On 27 December, Wilson and the Coxswain made it back to the ship, having crossed the island on foot, without food for the past two days. They reported the location of the Master and the whaleboat, and as the weather had eased FitzRoy set out in another boat with a week's supplies for them, only to meet the whaleboat making its way back before he had lost sight of his ship. Having given them food and two sailors to help them row, he decided that as the weather was fine he would continue 'to the place they had left, in order to do what the bad weather had prevented the master from doing'. Mission accomplished, he returned to the ship the next day. It was only then that he learned that as Wilson and the Coxswain had been making their way down to the shore to attract the attention of those on the *Beagle*, 'the Fuegians took advantage of their weak state to beat the coxswain and take away some of his clothes'. By then, it was too late to do much about this, but FitzRoy took a party ashore to search for the natives responsible. He was met with a show of force from the natives, a group which included only eight men, but who armed themselves with clubs, spears and 'swords, which seemed to have been made out of iron hoops, or else were old cutlasses worn very thin by frequent cleaning'. Nothing more came of the incident, and the ship sailed that same day; but it could no longer be assumed that awe of the white man would be enough to keep the Fuegians in their place.

The work proceeded without further drama for the next month. It's a sign of just how thorough FitzRoy's survey was that on 12 January 1830, for example, having identified a particular bay as a potential safe haven for ships rounding the Cape, he describes taking bearings from the summits of two different mountains, while Stokes and FitzRoy between them determined the latitude eight times using four different sextants, 'and as they all agreed, within fifteen seconds, I supposed their mean to be nearly correct'. In view of this outstanding attention to detail, it is no real surprise that FitzRoy's charts were still in use, as the main source of information about the waters around Tierra del Fuego, well over a hundred years later, even after the Second World War. A highlight of the voyage came a couple of days later, when *Beagle* entered the Barbara Channel, passing through a region surveyed by Skyring in the *Adelaide*, and linking the two surveys together; they delighted in taking note of the various capes and mountains named by him, and easily recognisable from his sketches. Just beyond the limit of Skyring's survey, FitzRoy spotted a high mountain which he gave the name St Paul's, from a fancied resemblance to the dome of that cathedral. As well as carrying out his assigned duties diligently, FitzRoy was always looking out for future possibilities. After describing the appearance of rock with a sulphurous smell, which he suspects of containing metal ores, he regrets that there is nobody on board the *Beagle* with the skill to identify it, and determines that: 'If ever I left England again on a similar expedition, I would endeavour to carry out a person qualified to examine the land; while the officers, and myself, would attend to hydrography.' Those prophetic words were written on 24 January 1830; within a few days, the chain of events that would soon lead to the prophecy coming true was set in action.

On 29 January, with the *Beagle* in safe anchorage, the Master was dispatched in the whaleboat once again, to survey a headland which Captain James Cook had named Cape Desolation. The whaleboat was a fine new one, built at San Carlos by Jonathan May, the *Beagle*'s carpenter, to replace the one washed away; FitzRoy had no worries for the wellbeing of its crew even when the weather yet again turned for the worse. As the wind began to blow a gale, FitzRoy's only concern was that 'however safe a cove Mr. Murray might have found, his time, I knew, must be passing most irksomely, as he could not have moved about since the day he left us'. By 4 February, the weather having eased, FitzRoy was surprised that the Master had not yet returned, but 'did not feel much anxiety, but supposed he was staying to take the necessary angles and observations, in which he had been delayed by the very bad weather'.

At three o'clock on the morning of 5 February, however, FitzRoy had a rude awakening. He learned that the whaleboat had been stolen by natives, and that the Coxswain and two men had just reached the ship with the news, in 'a clumsy canoe, made like a large basket, of wickerwork covered with pieces of canvas, and lined with clay, very leaky, and difficult to paddle'.[10] The Master and the others were awaiting rescue at Cape Desolation, where 'their provisions were all consumed, two-thirds having been stolen with the boat, and the return of the natives, to plunder, and perhaps kill them, was expected daily'.

FitzRoy set out in a second whaleboat (also built by May in San Carlos) with provisions for eleven men for two weeks, planning to go in search of the stolen boat. He found Murray's party intact but anxiously awaiting his arrival. They were in a desolate cove on what had seemed a totally uninhabited island, and although the whaleboat had been well secured, they had not kept watch during the night, thinking that not even Fuegians would live in so isolated a spot. FitzRoy blamed nobody for this, handsomely stating in the *Narrative* that he would have made the same mistake himself, not suspecting that Fuegians would be found 'on this exposed and sea-beaten island'. Nevertheless, on this exposed and sea-beaten island, without provisions and in fear of their lives, Murray and his companions had carried out all the observations required for fixing the precise geographical location of the place, so the party, made up to eleven including both rescuers and rescued, set off in the five-oared whaleboat to search for the missing boat.

FitzRoy pursued the search with stubborn tenacity, first for ten days and then on a second expedition after returning to the *Beagle* to replenish. He excused this obsession in the *Narrative* by pointing out that the surveying work was being carried out from the *Beagle* anyway, and that the information he gained about the Fuegians was important in its own right; but there is more than a hint here of an unwillingness to accept defeat and cut his losses, part of the characteristic FitzRoy conviction that he always knew the best course of action to take (whatever his orders might say) and would pursue it to the bitter end regardless of what others thought – not that there could be any possibility of anyone on the *Beagle* objecting to whatever course of action their Captain might choose to follow.

It's also true that to some extent FitzRoy was drawn into the extended search by initially seeming to be hot on the heels of the stolen boat. The thought that it might be just around the corner led him round one corner after another, until the fortnight had nearly passed. The very first place the searchers went to, a small island just two miles from the site of the theft, encouraged their efforts, because there they found the mast of the missing whaleboat, part of which had been cut off, clearly using an axe that had been in the boat. The island lay in a bay dotted with islands large and small, and FitzRoy 'resolved to trace the confines of the bay, from the west, towards the north and east, thinking it probable that the thieves would hasten to some secure cove, at a distance, rather than remain upon an outlying island, where their retreat might be cut off'.

Undaunted by gale force winds that brought hail and rain, they pursued the search, and on 7 February, thirty miles from Cape Desolation, came across a native family in two canoes. A search of the canoes revealed the lead line[11] from the whaleboat, and FitzRoy immediately took the owner of that canoe into his own boat, 'making him comprehend that he must show us where the people were, from whom he got it'. The Fuegian directed them to a small cove, where they found a party of women and children, with just one old man and a youth of about eighteen, and several items from the missing whaleboat, including the axe and the boat's tool bag. Convinced that the men from the families must be out sealing in the whaleboat, FitzRoy set off in pursuit of them, now with two native guides, his original captive and the youth from the camp, who seemed happy to accompany them. But FitzRoy was genuinely puzzled to know how to treat the miscreants: 'we had always behaved kindly to the Fuegians wherever we met them, and did not yet know how to treat them as they deserved, although they had robbed us of so great a treasure'. Rather than punish the natives in the camp, they only took back their own things; and the two men taken as guides were given some clothes and red caps, with which they were very pleased. But the various kindnesses were not to be repaid.

That night, the guides were not tied up as prisoners, but allowed to sleep by the fire with the men from the boat; in the morning, they were gone, taking with them two tarpaulin coats, which Mr Murray had put over them to keep them warm, although 'treated as he had so lately been, one might have thought he would not have been the first to care for their comfort'. Hoping to obtain another guide, FitzRoy headed back to the camp where he had found equipment from the whaleboat, but all the natives ran away at their approach. Now, for the first time FitzRoy took punitive action against the Fuegians, destroying their canoes to prevent them spreading warnings about the searchers or 'any intelligence likely to impede the return of our boat, which we daily expected'. After a fruitless search further around the shores of the bay, on 9 February they returned straight across the bay to Cape Desolation, then carried out a more detailed search of the region. FitzRoy's interest in the weather, and in particular the ability of the barometer to foretell bad weather, surfaces in his *Narrative*, where he writes on 11 February: 'The gale was extremely heavy . . . the barometer foretold it very well, falling more than I had previously seen.'

Next day, FitzRoy made a third visit to the place where he had found the equipment from the whaleboat. It had been abandoned, but late in the day, after a long search, he found the Fuegians in a cove some way along the coast. This time, he intended to take prisoners, 'as many as possible, to be kept as hostages for the return of our boat'. To that end early on the 13th the men from the *Beagle* encircled the Fuegian camp and closed in from all sides, almost reaching the camp before the dogs got wind of them and began barking:

At first the Indians began to run away; but hearing us shout on both sides, some tried to hide themselves, by squatting under the banks of a stream of water. The foremost of our party, Elsmore by name, in jumping across this stream, slipped, and fell in just where two men and a woman were concealed: they instantly attacked him, trying to hold him down and beat out his brains with stones; and before any one could assist him, he had received several severe blows, and one eye was almost destroyed, by a dangerous stroke near the temple. Mr. Murray, seeing the man's danger, fired at one of the Fuegians, who staggered back and let Elsmore escape; but immediately recovering himself, picked up stones from the bed of the stream, or was supplied with them by those who stood close to him, and threw them from each hand with astonishing force and precision. His first stone struck the master with much force, broke a powder-horn hung round his neck, and nearly knocked him backwards: and two others were thrown so truly at the heads of those nearest him, that they barely saved themselves by dropping down. All this passed in a few seconds, so quick was he with each hand: but, poor fellow, it was his last struggle; unfortunately he was mortally wounded, and, throwing one more stone, he fell against the bank and expired . . . That a life should have been lost in the struggle, I lament deeply; but if the Fuegian had not been shot at that moment, his next blow might have killed Elsmore.

The upshot of the struggle was that FitzRoy had eleven captives – two men, three women and six children. They seemed eager to help, and pointed out the direction in which they said the whaleboat had gone – the opposite to the direction pointed out by the earlier 'guides'. But FitzRoy was learning not to trust the Fuegians an inch. He decided to return to the *Beagle*, shift the position of the ship to the east, then carry out another search by boat, 'carrying some of my prisoners as guides, and leaving the rest on board to ensure the former remaining, and not deceiving us'. The overloaded whaleboat got back to the *Beagle* on 15 February, where the Fuegians were fed with fat pork and shellfish 'which they liked better than any thing else', and clothed with old blankets. Much later, FitzRoy learned that the captives were indeed members of the very family that had stolen the whaleboat from the Master.

After moving the ship to another safe anchorage, on 17 February FitzRoy set out once again, this time taking both a whaleboat and the cutter, and a week's provisions. He took a young Fuegian man in the whaleboat as a guide, and the Master carried two women in the cutter, on the natural assumption that since their children were hostages on the *Beagle*, they would behave themselves. 'They appeared to understand perfectly that their safety and future freedom depended upon their showing us where to find the boat.' But on the first night, the Fuegians slipped away in the dark, leaving their piles of blankets undisturbed behind them, like empty chrysalises. The two boats separated and searched the region independently, but found scarcely a trace of the stolen boat – just the sleeve of Mr Murray's tarpaulin coat in one canoe. And when he returned to the *Beagle* on the evening of 23 February, he found that all the prisoners on board, except three children, had escaped by swimming ashore the previous night. 'Thus, after much trouble and anxiety, much valuable time lost, and as fine a boat of her kind as ever was seen being stolen from us by these savages, I found myself with three young children to take care of, and no prospect whatever of recovering the boat.' But still, during FitzRoy's absence Stokes had made detailed plans of the region, and 'this cruise had also given me more insight into the real character of the Fuegians, than I had then acquired by other means, and gave us all a severe warning which might prove useful at a future day, when among more numerous tribes who would not be contented with a boat alone'.

FitzRoy became convinced by this incident of the need to establish proper communications with the Fuegians, either by learning their language or by teaching some of them English. But the first necessity was to replace the lost boat. FitzRoy had established that the best way to carry out his surveying was to anchor the ship safely and send out one boat in each direction, and the cutter required too many men and was not so handy as a whaleboat, while the ship's small boat 'was only fit for harbour duty'. With great reluctance, the best spare spar that the *Beagle* carried, intended to replace any damaged lower mast, was sacrificed to the carpenter's saw, and Jonathan May built yet another whaleboat for the ship. Late on 28 February, *Beagle* anchored in Christmas Sound, the place chosen by FitzRoy to carry out the boat-building. This was, as he put it, 'the very spot' where Captain Cook had stopped on his voyage around the world, beneath a prominent feature which Cook had called York Minster, after the resemblance of its outline to the shape of that building. Next day, because the anchorage turned out to be rather exposed, they shifted the ship into a more sheltered spot, with a good supply of wood and water, and the boats were sent out to carry on surveying while the carpenter set about his work. On this occasion, the Master took the cutter, and carried with him two of the Fuegian children, with instructions to hand them over to any natives he might find;[12] 'the third, who was about eight years old, was still with us: she seemed to be so happy and healthy, that I determined to detain her as hostage for the stolen boat, and try to teach her English'. The little girl was given the name 'Fuegia Basket' by the sailors, in reference to the makeshift 'canoe' which had brought the news of the loss of the whaleboat to the *Beagle*.

While the new whaleboat was being built, on 3 March the *Beagle* was visited by more Fuegians, who seemed eager to come aboard. By now, though, FitzRoy's attitude to the natives had changed: 'I had no wish for their company, and was sorry to see that they had found us out;

for it was only to be expected that they would soon pay us nightly as well as daily visits, and steal every thing left within their reach.' Gestures failed to drive the Fuegians away, and a pistol fired over their heads only shifted them as far as a nearby point of land. FitzRoy's immediate reaction was to send a boat to drive them off. But then he had another idea. Reflecting that:

> By getting one of these natives on board, there would be a chance of his learning enough English to be an interpreter, and that by his means we might recover our lost boat, I resolved to take the youngest man on board, as he, in all probability, had less strong ties to bind him to his people than others who were older, and might have families. With these ideas I went after them, and hauling their canoe alongside of my boat, told a young man to come into it; he did so, quite unconcernedly, and sat down, apparently contented and at his ease. The others said nothing, either to me or to him, but paddled out of the harbour as fast as they could.

So FitzRoy acquired (or kidnapped) a second Fuegian, who soon became known as York Minster, after the promontory overlooking the bay where he was caught. Although sullen at first, 'York' cheered up as soon as he was cleaned and clothed, and it was made clear to him that he could move about the ship wherever he liked.

Five days later, there was a much more serious encounter. Two canoes full of men were seen from the ship, heading for the shore, and when FitzRoy went after them in a small boat with just two men, near where the carpenters were at work he found a party of a dozen men and just two women preparing their camp. They 'were getting very bold and threatening in their manner' when Lieutenant Kempe and six men arrived, just in time to rescue FitzRoy from an attack. It seemed that this was a party prepared for violent action, 'being much painted, wearing white bands on their heads, carrying their slings and spears, and having left all their children and dogs, with most of their women, in some other place'. He surmised that they had seen the cutter heading west full of men, and, supposing that not many men had been left on board the *Beagle*, were planning to attack the ship herself. Having gone back with Kempe to fetch reinforcements, FitzRoy took an armed party in two boats to confront the Fuegians, who left their canoes at the foot of a rock rising steeply from the water, and bombarded the boats with accurately thrown stones, without preventing the canoes from being captured. Shots were fired in return, but nobody was hit, and the only significant casualty in the boats was a seaman stunned by a stone. The canoes were found to contain empty bottles from the ship's stores (they had originally contained beer), and other items from the stolen whaleboat. The Fuegians escaped into the bushes.

Next day, convinced that he was once again hot on the heels of the missing boat, FitzRoy took a party ashore to investigate smoke coming from a nearby island, and found an encampment from which the natives had fled at their approach. In one of the abandoned shelters, they found a distinctive piece of light rope, a kind known as King's white line, brand new and clearly part of the booty from the whaleboat. While the shore party were searching for the natives from the encampment, two canoes suddenly shot away from the island, paddling hard into a rough sea; FitzRoy gave chase and caught one of the canoes, from which a young man and a girl jumped overboard to swim for shore, abandoning an old woman and a child. FitzRoy's instant decision was to leave the canoe and the swimming girl, and to capture the young man, although it took a quarter of an hour to get him in to the boat and subdue him. As well as providing FitzRoy with another potential interpreter – who they named 'Boat Memory' since he obviously knew all about the missing whaleboat – this provided FitzRoy with more justification, if he needed any, for his course of action:

> They are a brave, hardy race, and fight to the last struggle; though in the manner of a wild beast, it must be owned, else they would not, when excited, defy a whole boat's crew, and, single-handed, try to kill the men; as I have witnessed. That kindness towards these beings, and good treatment of them, is as yet useless, I almost think, both from my own experience and from much that I have heard of their conduct to sealing vessels. Until a mutual

understanding can be established, moral fear is the only means by which they can be kept peaceable.

So FitzRoy took it upon himself to establish that mutual understanding, not just for the benefit of Europeans, but, as any of his Victorian contemporaries[13] would have understood without comment, for the natives' own good.

On 14 March, Murray returned with the cutter, and on the 15th, 'raining and blowing: – as usual I might say', FitzRoy went off on another surveying expedition while the whaleboat was being completed. Returning on the 30th, he found that all was well on board the *Beagle*, there had been no further visits from the Fuegians, and a fine new whaleboat had been completed on the 23rd. With nothing left to detain them, the ship left March Harbour, 'so called from our having passed the month of March in it', on the 31st, and headed eastward around the bottom of Tierra del Fuego. From a surveying point of view, the highlight of this part of the voyage was the discovery (by Mr Murray) of the Beagle Channel, an almost canal-like cut that slices west–east through the southern tip of Tierra del Fuego. But FitzRoy's hopes of establishing mutual understanding with the Fuegians took a knock when the ship came across different tribes who did not even speak the same language as his captives. At first, 'much enmity appeared to exist between them', and ' "York" and "Boat" would not go near them; but afterwards took delight in trying to cheat them out of the things they offered to barter; and mocked their way of speaking and laughing; pointing at them and calling them "Yapoo, yapoo".' So FitzRoy dubbed the tribes from this part of Tierra del Fuego 'Yapoos'.

On 18 April, the *Beagle* reached a significant landmark, anchoring in a bay near the tip of South America, Cape Horn itself. On the 19th, FitzRoy and a boat party landed on Horn Island, and on the 20th they climbed to the peak of Cape Horn, where they carried out surveying observations, left a message in a glass jar buried under a mound of stones eight feet high, raised the Union Flag over the mound, and drank a toast to the health of King George IV.[14] There was plenty to celebrate; from now on, they would be travelling northward, up the eastern seaboard of South America to Rio, and then on home to England.

There was still surveying work to be done, and still foul weather to endure. And still time for FitzRoy to add another native to his collection. On 11 May, on another surveying trip in the boats, FitzRoy's party came across some natives in three canoes, eager to barter.

> We gave them a few beads and buttons, for some fish; and, without any previous intention, I told one of the boys in a canoe to come into our boat, and gave the man who was with him a large shining mother-of-pearl button ... 'Jemmy Button', as the boat's crew called him, on account of his price, seemed to be pleased at his change.

But why acquire yet another native, when there was no longer any hope of recovering the long-lost whaleboat, and no prospect of going back around Cape Horn to return the three he already had to the homelands? Although FitzRoy tells us that he acted on impulse, his subconscious mind at least must have already been considering what to do with the natives. But it is only in his *Narrative* entry for 10 June, when the ship, low on provisions and certain to be late for the rendezvous with King in Rio on the 20th, was about to leave Tierra del Fuego entirely that he confides that:

> I had previously made up my mind to carry the Fuegians, whom we had with us, to England; trusting that the ultimate benefits arising from their acquaintance with our habits and language, would make up for the temporary separation from their own country. But this decision was not contemplated when I first took them on board; I then only thought of detaining them while we were on their coasts; yet afterwards finding that they were happy and in good health, I began to think of the various advantages which might result to them and their countrymen, as well as to us, by taking them to England, educating them there as far as might be practicable, and then bringing them back to Tierra del

Fuego. . . . I incurred a deep responsibility, but was fully aware of what I was undertaking.

The Fuegians were much slower in learning English than I expected from their quickness in mimickry, but they understood clearly when we left the coast that they would return to their country at a future time, with iron, tools, clothes, and knowledge which they might spread among their countrymen. They helped the crew whenever required; were extremely tractable and good-humoured, even taking pains to walk properly, and get over the crouching posture of their countrymen.

It was actually 2 August before the *Beagle* reached Rio de Janeiro – fully six weeks late, but this seems to have been accepted as a justifiable delay in view of the nature of the work she had been doing and the quality of FitzRoy's charts. Almost immediately, FitzRoy learned that his father had died, at the age of sixty-five, on 20 December 1829 – shortly before the series of events that had led to the Fuegians being on board the *Beagle*. This came as a shattering blow, as we shall shortly see, and increased his impatience to be home; departure, happily, would not be long delayed, but the voyage, less happily, was to be a long one. On 6 August, *Adventure* and *Beagle* set off for England together, leaving the *Adelaide* to act as tender for the flagship.

The journey home is described in the *Narrative* as 'a most tedious passage'; it took until 14 October, largely because King had to return home via the Cape Verde islands, on the other side of the Atlantic (we do not know why this detour was considered essential; FitzRoy merely says that it was 'for a particular purpose'). This meant heading east from Brazil, crossing the equator far to the east then heading north to England – all against the prevailing winds – instead of heading first north across the equator and then east, with the prevailing winds. But for the Fuegians at least the voyage was full of interest, rounded off when a steamship passed the *Beagle* at night, on its way into Falmouth Harbour.

What extraordinary monster it was, they could not imagine. Whether it was a huge fish, a land animal, or the devil (of whom they have a notion in their country), they could not decide; neither could they understand the attempted explanations of our sailors, who tried to make them comprehend its nature: but, indeed, I think that no one who remembers standing, for the first time, near a railway, and witnessing the rapid approach of a steam-engine, with its attached train of carriages, as it dashed along, smoking and snorting, will be surprised at the effect which a large steam-ship, passing at full speed near the Beagle, in a dark night, must have had on these ignorant, though rather intelligent barbarians.

FitzRoy's joy at returning home was tempered by the loss of his father. One letter, written to Fanny over a period of several days at the end of September and beginning of October 1830, and sent from Falmouth, shows the depth of FitzRoy's response to the news, and gives us a rare insight into his feelings:

I do not wonder at *your* not writing to me lately, for I know what your feelings have been as well as if I had been close to you all the time. – You and I were always more alike in our dispositions than either one of us to others, and knowing you by that rule I feel certain that the fatal loss we have last sustained has been felt by you, more than any one, even George, and yet I question whether you did not bear the shock far better than he or others – I hope so, and I hope that by this time you have partly reconciled your mind to a calamity which was inevitable, yet one might have hoped, would have been postponed for many long years . . .

You, and you only know how I valued his slightest word, for, like yourself it is not natural to me to be much, or suddenly affected, outwardly, by what occurs. I have been working hard, Fanny, and have run many risks during the last two years, & through all, I have been influenced by the thought that it would give him satisfaction. His approbation I looked to as the true reward of any hard times I might pass, and I thought little of any other person's. I arrived at Rio, happy in the thought of having earned his approval, happy in the idea that he would welcome me back more gladly than he ever did before, and in the first letter I opened I found myself stamped upon, for the sudden reverse gave me that kind of sensation . . . but I did not shut myself up, nor neglect my duty for a minute, – I found that the more I employed myself and forced occupation, the easier I got through the day. – My worst time was when alone and unemployed . . . What a life this is – the pains are far greater than the pleasures – and yet people set such a value upon existence, as if they were always happy.

This glimpse into FitzRoy's soul confirms much that we might have guessed from the outside – the character of a little boy who lost his mother, was sent away to school and then to sea, worshipped a distant father and felt he could never be good enough for that father's approval. A man who kept his feelings to himself, was driven by a strong sense of duty, and found hard work the best remedy in times of despair. But there was much more to FitzRoy than this.

In the latter part of the letter, written several days later, we hear the voice of the old enthusiastic FitzRoy, enthusing about the voyage and looking to the future:

The Beagle has been very fortunate indeed, she has been in many difficulties, and is here to answer for herself, unhurt. She is a very fine little vessel indeed, and her crew, thorough English seamen . . . How are your childers? Pray write me a few lines, and tell me about them and yourself. Is George married? If to Louisa Harris, I am heartily glad, & wish him joy of an excellent wife and you, of an amiable friend. God bless you my dear Sister.

George had indeed married Louisa Harris, on 30 July that year, helping to ensure that in spite of the loss of their father there was a warm family welcome awaiting FitzRoy in England; the moral couldn't have been clearer – life goes on. But he had a 'family' of another kind to settle before he could relax and take a rest himself, and plenty of hard work to keep him occupied if resting brought back the depression he had experienced in Rio.

CHAPTER THREE

Interlude in England

THE *BEAGLE* WAS so worn out at the end of her extensive voyage that she was in need of a complete refit – practically a rebuilding – but as she was paid off at Plymouth and turned over to the dockyard (she was officially decommissioned on 27 October, with the crew who had been through so much together dispersed to other duties), that didn't seem to be something FitzRoy would have to worry about. He had other things on his mind.

While on the way home, FitzRoy had written a long letter to King, explaining how the Fuegians came to be on board the *Beagle* and spelling out his hopes and plans for their future – to educate them in civilised ways and 'after two or three years' to 'send or take them back to their countries, with as large a stock as I can collect of those articles most useful to them, and most likely to improve the condition of their countrymen, who are now scarcely superior to the brute creation'. The letter was forwarded by King to the Admiralty, which elicited the generous response that 'their Lordships will not interfere with Commander Fitz-Roy's personal superintendence of, or benevolent intentions towards these four people, but they will afford him any facilities towards maintaining and educating them in England, and will give them a passage home again'.

The Fuegians were sent to a quiet farmhouse, a few miles inland, where FitzRoy hoped that they would be free not only from the kind of intrusive curiosity they would arouse in the port, but also from the risk of contagion by infectious diseases. Unfortunately, in spite of them being vaccinated for smallpox, Boat Memory, seemingly the most intelligent of the four and FitzRoy's favourite, was taken ill with symptoms of the disease. Early in November, FitzRoy had to make his first call on the promise made by the Admiralty to provide facilities for the natives, and they were all taken in to the Royal Hospital at Plymouth, accompanied by James Bennett, the former coxswain of the *Beagle*.[1] FitzRoy had to leave them in Plymouth to travel to London to report on his surveying mission, and it was there that he received news that, in spite of receiving the best treatment that the doctors could provide, Boat Memory had died. 'It may readily be supposed that this was a severe blow to me, for I was deeply sensible of the responsibility which had been incurred; and, however unintentionally, could not but feel how much I was implicated in shortening his existence.' But the vaccinations of the other three Fuegians were successful, and none of them caught the disease.

Alongside his still extensive and time-consuming duties in London concerning the survey, FitzRoy now had to find a longer term home for his charges. He turned to the Church Missionary Society, establishing contacts that would have a profound influence on the later course of his life. There is nothing to suggest, however, that at this time FitzRoy was unusually devout, or committed to a deeply fundamentalist, literal interpretation of the Bible – he was just an ordinary Christian going about his duties in an appropriate way. Indeed, as he wrote near the end of volume two of the *Narrative*:

I suffered much anxiety in former years from a disposition to doubt, if not disbelieve, the inspired History written by Moses. I knew so little of that record, or of the intimate manner in which the Old Testament is connected with the New, that I fancied some events there related might be mythological or fabulous, while I sincerely believed the truth of others; a wavering between opinions, which could only be productive of an unsettled, and therefore unhappy, state of mind.

It seems, indeed, to have been a result of long hours reading the Bible and reflecting on its contents during his second voyage in command of the *Beagle* that convinced FitzRoy of the literal truth of every word in it; this belief was reinforced by his marriage to a woman who held fundamentalist Christian views, and it may be that his Bible reading on the second voyage was a result of his engagement to her.[2] That is getting ahead of our story, but it is worth emphasising here that there is no evidence to support the supposition sometimes made that from the outset of the second voyage FitzRoy was hoping to find geological evidence for Biblical events, in particular the Flood.

The upshot of FitzRoy's approach to the Church Missionary Society was that the Reverend William Wilson, of Walthamstow, took the Fuegians into his parish, where they lived with the master of the infants' school, and studied with his pupils. The schoolmaster and his wife were 'pleased to find the future inmates of their house very well disposed, quiet, and cleanly people; instead of fierce and dirty savages'. They stayed there from December 1830 to October 1831, being taught English and 'the plainer truths of Christianity', as well as 'the use of common tools, a slight acquaintance with husbandry, [and] gardening'. Jemmy Button (about fourteen years old) and Fuegia Basket proved good pupils, but York Minster, who was much older than them, probably in his mid twenties, 'was hard to teach, except mechanically'. The strangers naturally aroused considerable curiosity, and were often visited by people from the neighbourhood and further afield, who frequently gave them presents; a particular benefactress was FitzRoy's sister, who they often talked about, even after they had left England, referring to her as 'Cappen Sisser'. Towards the end of their stay in England, in the late summer of 1831, the King (now the sixty-five-year-old William IV, the brother of George IV, who had died in June 1830) invited the Fuegians, accompanied by FitzRoy, to meet him at St James's, where Queen Adelaide was also present and gave Fuegia one of her own bonnets and one of her own rings, as well as giving FitzRoy money to buy the little girl clothes 'when she should leave England'; by then, less than a year after the *Beagle* had returned from South America, plans for the departure of Fuegia and her two companions were already well advanced.

FitzRoy, having completed his work in connection with the survey in March, had been disappointed to learn that although the survey of southern South America was far from complete, the Lords of the Admiralty had no immediate plans to continue the work, and that in consequence there was little likelihood of a Royal Navy ship visiting Tierra del Fuego in the next couple of years. In June, 'having no hopes of a man-of-war being sent to Tierra del Fuego', and unwilling to trust the Fuegians to the care of any other kind of vessel without himself, he entered into a contract with the owner of a small merchant ship, the *John of London*, to carry himself and 'five other persons to such places in South America as I wished to visit'. He had permission from the Admiralty for twelve months' leave of absence, and the arrangements were well in hand (at considerable expense to FitzRoy) when 'a kind uncle, to whom I mentioned my plan, went to the Admiralty, and soon afterwards told me that I should be appointed to the command of the *Chanticleer*, to go to Tierra del Fuego'. Never was 'interest' more blatantly at work! Although we are not told which uncle carried out the act of nepotism, it must have been either the then Duke of Grafton or (more probably in the light of future events; see Chapter Six) the Earl of Londonderry (the brother of 'our' Lord Castlereagh). But FitzRoy also had another valuable ally, who, together with his uncle, played a big part in persuading the Lords of the Admiralty of the desirability of another South American survey. This was Captain Francis Beaufort, of wind scale fame, who was at that time Hydrographer to the Admiralty.

Beaufort, who was born in County Meath, Ireland, in 1774, was the son of the Rector of Navan, and through his father's connections had gone on to the books of the Royal Navy ship *Colossus* at the age of thirteen, although he only went to sea (and then initially with the East India Company, not the Royal Navy) in 1789; he had begun keeping a weather journal the next year, in 1790, when he joined the frigate HMS *Latona* as a Midshipman. In 1794, now serving on the *Aquilon*, he was present at the battle known as the 'Glorious First of June', where the British inflicted a famous and heavy defeat on the French, and by 1796, at the age of twenty-one (just two weeks short of his twenty-second birthday) he was a Lieutenant serving on board the frigate *Phaeton*. *Phaeton*'s successes over the next few years brought in enough prize money to make even her junior Lieutenant a comparatively wealthy man, but he paid the price in the autumn of 1800, when he was severely wounded in an encounter with French privateers. After a long spell ashore on half-pay, it was only in 1805 (the year FitzRoy was born) that he was given his first command, HMS *Woolwich*, in which he carried out a hydrographic survey of the River Plate region of South America. It was during this period that he developed the early version of his wind scale system and his weather notation, a shorthand way of noting down the salient features of the weather.

Made Post[3] in 1810, by 1812 Beaufort was involved in a hydrographic survey of the Eastern Mediterranean, which was combined with patrol duties against the pirates that plagued the region. In June 1812, a shore party making astronomical observations was attacked by hostile locals, and Beaufort himself led the party that went to their rescue; while they were rowing back to his ship, the *Fredrikssteen*, Beaufort was hit by a bullet from the shore which fractured his femur, close to the hip. After several months' convalescing, he returned home to shore duties, which eventually led him to take up the post of Hydrographer in 1829, while FitzRoy was in South America. He didn't leave the post until 1855, when he was eighty-one, after sixty-eight years' continuous service in the Navy.

Beaufort was just the man to appreciate FitzRoy's qualities as a seaman and surveyor, and the two also hit it off personally, with FitzRoy becoming something of a protégé of the older man. If interest was at work in persuading their Lordships that a second survey of Tierra del Fuego was desirable, there could be no doubt in anyone's mind, least of all the Hydrographer's, about the right man to command such an expedition. Captain King had retired, and in his report to the Admiralty[4] had been fulsome in his praise of FitzRoy, stating that he had been fortunate that Sir Robert Otway had placed FitzRoy in command of the *Beagle*, referring to the hardships and difficulties of the survey work undertaken in the open boats and the splendid example provided by FitzRoy, and concluding by informing their Lordships that the young Commander 'merits their distinction and patronage', recommending him 'in the strongest manner to their favourable consideration'.

On examination, the *Chanticleer* was found to be unfit for the duties required, but a happy alternative was available. FitzRoy was left to pay the best part of a thousand pounds to get out of his commitment to the owner of the *John of London*, but on 27 June he was formally reappointed to the command of the *Beagle* with instructions to prepare her for a second voyage to South America. On the same day, 'two of my most esteemed friends, Lieutenants Wickham and Sulivan, were also appointed', doubtless at FitzRoy's request. They were the first of many old companions who would also return to the ship, although knowing full well the conditions they would be sailing into – itself a striking indication of FitzRoy's popularity and ability as a commander.

At the request of the Church Missionary Society, the Admiralty agreed that two missionaries should accompany the Fuegians back to their homeland, to live among the natives in the hope of establishing a permanent mission settlement; FitzRoy's orders were drafted to allow him to revisit the missionaries before finishing his surveying work, to offer any advice or assistance they needed, or to rescue them if necessary. In fact (and hardly surprisingly) only one young man willing to volunteer for this hazardous experience could be found before the *Beagle* left England.

She had been officially commissioned on 4 July 1831, and immediately taken into dock for the complete overhaul required before she would be fit once again for long service far from home. No expense seems to have been spared in preparing her for the voyage, with a considerable amount of that expense coming from FitzRoy's own pocket, for items that he considered essential but which the Admiralty regarded as luxuries. Not that the Admiralty got off lightly – the official cost of the refit, £7,583, was only £221 less than the original cost of building the *Beagle*. The ship was effectively rebuilt, under the close supervision of FitzRoy and incorporating at his insistence the latest nautical technology, with the upper deck raised eight inches at the stern and twelve forward, which not only gave more room below but also improved her handling at sea. The arrangement of the tiny cabins was altered in the rebuilding, providing a more convenient layout (as far as anything could be convenient in a ship that size), and having the bonus that this meant that FitzRoy would no longer be sleeping in the very space where Stokes had first shot himself then suffered a lingering and painful death. A sheath of two-inch-thick fir planking was nailed to the bottom of the *Beagle* and covered with new copper, increasing the 'burden' of the ship from 235 tons to 242 tons; the rudder was fitted in a more modern and efficient way; the capstan was replaced by a windlass; lightning conductors were fitted to all the masts (for the first time in a Royal Navy ship); and the ropes, sails and spars 'were the best that could be procured'. The arrangements, said FitzRoy, 'left nothing to be desired', and:

> Never, I believe, did a vessel leave England better provided, or fitted for the service she was destined to perform, and for the health and comfort of her crew, than the Beagle. If we did want any thing which could have been carried, it was our own fault; for all that was asked for, from the Dockyard, Victualling Department, Navy Board, or Admiralty, was granted.

Still rigged as a barque, the *Beagle* had unusually strong masts and rigging for a vessel of her tonnage, with chains rather than ropes used wherever practicable. But it was in boats that *Beagle* really excelled. As well as a dinghy carried astern, there were no fewer than four whaleboats (two of them FitzRoy's personal property); two of them, each 28 feet long, were carried on skids over the quarterdeck, and two, each 25 feet long, on the quarters of the ship; in addition, two more boats, stowed one inside the other, were secured diagonally amidships across the deck. And remember that *Beagle* herself was only 90 feet long and 24 feet 6 inches wide (less than the length of one-and-a-half cricket pitches!). It's a wonder the sailors had room to move about on deck, but this was nothing compared to the crowded conditions below decks, where even FitzRoy described the cabins for the officers as 'extremely small', joking that they were 'filled in inverse proportion to their size'.

One reason for the overcrowding was the sheer amount of stores required for the voyage; the other, of course, was the number of extra people, supernumeraries, that the ship carried. Richard Matthews, the young missionary, and the three Fuegians arrived from Walthamstow in October (on board a steamship, which must have delighted the Fuegians), bringing with them large quantities of 'clothes, tools, crockery-ware, books and various things which the families at Walthamstow and other kind-hearted persons had given'. All of the gifts were well-intentioned, and much was of practical value; but the mirth of the sailors was thoroughly aroused by the presence of several complete sets of crockery – dinner services, tea services and the like – all of which had to be carefully stowed in the crowded holds of the *Beagle*. Many of those sailors had been with FitzRoy on the previous voyage, and knew only too well the futility of expecting the Fuegians to take tea with the missionary. Matthews himself hardly inspired confidence, and FitzRoy's efforts to find something nice to say about him are clearly forced:

Although he was rather too young, and less experienced than might have been wished, his character and conduct had been such as to give very fair grounds for anticipating that he would, at least, sincerely endeavour to do his utmost in a situation so difficult and trying as that for which he volunteered.

FitzRoy also engaged (at his own expense, £300, although the Admiralty agreed he could be fed from the ship's rations) an artist, Augustus Earle, to record the sights of South America; and with no fewer than twenty-two chronometers being carried (in FitzRoy's own tiny cabin) to ensure accurate determination of longitude in a chain right around the world, he employed George Stebbing, the eldest son of the mathematical instrument maker at Portsmouth, as his private assistant to look after the instruments. The chronometers were yet another expense largely borne by FitzRoy personally. The Lords of the

Admiralty were of the opinion that five would be sufficient for the purposes of the voyage, and made their views known after FitzRoy had already selected nine from the stores and taken them on board. Captain Beaufort passed on the news in a letter to FitzRoy in which he said:

> This grieves me exceedingly, but I must submit; choose therefore those you like best, and when you have made up your mind, return the rest into store. Or would you like me to write to the Board, to solicit your being allowed to retain seven out of the nine, on the ground of the accidents that may happen?[5]

But FitzRoy's reaction, typically, was to take the matter into his own hands, determined to do the job properly, even if it had to be at his own expense. Jumping off from an earlier letter in which Beaufort had suggested that eighteen chronometers 'might be enough' and that two more should be taken in case of accidents, he added another two for luck and purchased several additional instruments himself – for another £300 – while also persuading the manufacturers to lend a few more.

As well as his old friends Wickham and Sulivan, the officers on this voyage included John Lort Stokes, who had passed for Lieutenant but was as yet only rated Mate, and Midshipman Philip Gidley King, the fourteen-year-old son of the former captain of the *Adventure*;[6] the choice by Captain King of the *Beagle* for his son was, of course, another tribute to FitzRoy's ability. They also had the same carpenter as before, Jonathan May, and several other old shipmates. The total complement of officers and men (including six boys, a Sergeant of Marines and seven privates) was sixty-five, but the total number of people on board when the *Beagle* sailed from England at the end of 1831 was seventy-four. The nine supernumeraries were made up of the three Fuegians, Matthews, Earle and Stebbing, FitzRoy's steward, and Charles Darwin and his own servant.

Darwin was there for two reasons. First, FitzRoy was keeping true to his vow that if he ever returned to Tierra del Fuego, 'I would endeavour to carry out a person qualified to examine the land; while the officers, and myself, would attend to hydrography.' This is the way FitzRoy puts it in the *Narrative*:

Anxious that no opportunity of collecting useful information, during the voyage, should be lost; I proposed to the Hydrographer [Beaufort] that some well-educated and scientific person should be sought for who would willingly share such accommodations as I had to offer, in order to profit by the opportunity of visiting distant countries yet little known. Captain Beaufort approved of the suggestion, and wrote to Professor Peacock, of Cambridge, who consulted with a friend, Professor Henslow, and he named Mr. Charles Darwin, grandson of Dr. Darwin the poet,[7] as a young man of promising ability, extremely fond of geology, and indeed all branches of natural history. In consequence an offer was made to Mr. Darwin to be my guest on board, which he accepted conditionally; permission was obtained for his embarkation, and an order given by the Admiralty that he should be borne on the ship's books for provisions. The conditions asked by Mr. Darwin were, that he should be at liberty to leave the Beagle and retire from the Expedition when he thought proper, and that he should pay a fair share of the expenses of my table.

It seems clear, though, that FitzRoy had a second motive for inviting 'a person qualified to examine the land' along on the second voyage. This time, he would be in sole command, with no superior to consult or take orders from, and nobody he could share his thoughts with. He knew the strains of the voyage he was setting out on, and he also knew that he was subject to fits of depression. Although little of this comes through in his description of the first voyage, and there was no Darwin on that voyage to write about the mood-swings of the Captain, everything we learn about FitzRoy's character from Darwin and in later life confirms that this must have been the case on the first voyage as well. FitzRoy had an uncle who had killed himself in a fit of depression, and he had gained his step up into the *Beagle* thanks to the suicide of the previous captain – gloomy enough thoughts in the depths of the southern winter even if he had not been a depressive. Whatever else he wanted from the scientifically minded person he would carry with him, he wanted an intellectual and social equal, whose company would act as a safety valve, hopefully preventing FitzRoy from sinking too deeply into depression when the black mood struck him. Which is why Charles Darwin turned out to be the ideal man – we should say, the ideal gentleman – for the job, even though the route to his acceptance on board the *Beagle* was not as smooth as FitzRoy makes it seem.

CHAPTER FOUR

FitzRoy's Passenger

EVEN 'MERELY' AS a passenger on the *Beagle*, Darwin played such a large part in FitzRoy's life, and evolution played such a large part in Darwin's life, that it makes sense to set the scene for the voyage that was so important in both their lives by looking briefly at the earlier development of ideas about evolution. Charles Darwin did not invent the idea of evolution, which had developed steadily from the sixteenth century onwards as scientific learning developed in Europe. His important contribution was to offer an explanation of how evolution works – by natural selection. Even here, though, he was not unique, as we shall see. Darwin's rare value was in the breadth of his learning, the way in which he applied the scientific method to all his work, leaving no room for wishful thinking, and the clarity and beauty of his writing, which got his message across (and still gets his message across) to a wide audience. More than for any other great scientific advance, it is still true to say that the best way to learn about evolution is to read the original works of the scientist who put the idea forward.

What you might call evolutionary – although rather strange – ideas about how the living world got to be the way it is go back to the time of the Ancient Greeks; but we can pick up the story with Francis Bacon (1561–1626), who wrote in his book *Novum Organum*, published in 1620, about the way in which species vary naturally from one generation to the next. He pointed out that such natural variation could be used by the breeders of plants and animals to produce 'many rare and unusual results'. In Germany, Gottfried Leibniz (1646–1716), best remembered as a mathematician, was intrigued by fossils, and the possible relationship between the extinct ammonites and living species such as the nautilus. Speculating that the species had changed because the modern variety lived in a different environment from their ancestors (he did not actually use the word 'evolved'; the term evolution was not used in its modern biological context until 1826, by Robert Jameson, in Edinburgh), he wrote, 'it is credible that by means of such great changes [of habitat] even the species of animals are often changed'.

This was by no means an isolated view in the seventeenth century. In the eighteenth, Georges-Louis Leclerc, Comte de Buffon (1707–88), puzzled over the geographical distribution of similar but different species, and suggested that the North American bison might be descended from an ancestral variety of ox that had migrated there, where 'they received the impressions of the climate and in time became bisons'. Such ideas not only implicitly include the concept of evolution, but also the notion that species have evolved to fit their environments – adaptation.

The greatest of Darwin's predecessors as an evolutionary thinker, though, was his grandfather, Erasmus Darwin (1731–1802). A larger-than-life figure who was a doctor good enough to be asked by George III to become his personal physician (although he did not take up the post), a poet good enough to be considered a candidate for Poet Laureate (although he just missed out there) and a 'natural philosopher' good enough to be elected a Fellow of the Royal Society (although his most important piece of scientific work now bears somebody else's name), Erasmus Darwin combined several of his interests by writing about evolution in a long erotic poem, *The Botanic Garden*, as well as in a two-volume prose work, *Zoonomia*, published in 1794 and 1796, more than ten years before Charles Darwin was born.

Erasmus Darwin had a clear idea about the importance of evolution, but mistakenly thought that individual members of a species developed different characteristics during their lifetime, and these enhanced characteristics were passed on to their offspring. For example, if a deer needs to run away from a predator to avoid being killed, it will (on this picture; not really) somehow acquire a body slightly better adapted to running, more or less by willpower, and pass that slightly improved body plan on to the next generation through heredity. The same process, repeated over very many generations, then 'explains' why deer are such good runners today.

Curiously, almost exactly the same idea was developed independently in France by Jean Baptiste Lamarck (1744–1829), who published it in 1801. Lamarck was a full-time scientist, Professor of Zoology at the Museum of Natural History in Paris, and developed the idea much further, so it is perhaps just that it should be known today as 'Lamarckism' (it would certainly be confusing if it were called Darwinism!). What is less just is that it is often derided as a silly idea, when in fact it was a respectable attempt to put evolution on a scientific footing, and helped to stimulate discussion about the whole idea of evolution.

Lamarck's ideas were taken up by his disciple Geoffroy Saint-Hilaire Etienne (1772–1844), who not only promoted them but made the first clear statement of what we would now call the concept of survival of the fittest. Writing, in the 1830s, about the kind of modifications to an individual that Lamarck (and Erasmus Darwin) had described, he said: '. . . If these modifications lead to injurious effects, the animals which exhibit them perish and are replaced by others of a somewhat different form, a form changed so as to be adapted to the new environment.' Unfortunately, Saint-Hilaire also had some distinctly strange ideas about the relationships between species, and (among other things) tried to prove that the body plan of a mollusc is the same as that of a vertebrate. This led to all of his ideas being judged out of court, and by 1830 Lamarckism, tarred with the brush of Saint-Hilaire's dippy ideas, was no longer seen as respectable science in France. In England, hardly anyone knew of Erasmus Darwin's ideas about evolution. But young Charles Darwin was on the brink of his greatest adventure. It would be almost thirty years before the fruits of that adventure were presented to an astonished public. But although the public were astonished when Darwin's *Origin of Species* was published in 1859, the scientific world ought not to have been. Indeed, if it had not been for Saint-Hilaire's excesses the modern version of the theory of evolution by natural selection might well have been developed, from Lamarck's work, in France, while Charles was still at sea on board the *Beagle*.

We have already introduced one of Charles Darwin's grandparents, so we should flesh out his family background by introducing the others. Erasmus, as we have seen, was a successful country doctor, who lived near Shrewsbury, not far from the heartland of the Industrial Revolution in England. He became a gentleman amateur in science and literature, wealthy and influential in science, literature and society. He was also a huge man physically, with large appetites, addicted to life's pleasures, including food, wine and women. His first wife, Mary, died of drink in 1770 at the age of thirty, having produced five children; the third of these, Robert Waring Darwin (born in 1766) would become the father of Charles.

For the next ten years, Erasmus lived the high life (including fathering two children by the governess of his large brood), before, at the age of forty-nine, falling for a woman sixteen years his junior who was not only one of his patients but married to a rich man. Her husband died in 1780, Erasmus married the widow, Elizabeth, and together they shared the parental responsibilities for his five children by Mary, his two bastard offspring, and one bastard bequeathed by her late husband. Together, the couple produced seven more children before Erasmus died in 1802. Elizabeth outlasted him by thirty years.

One of Erasmus's great friends was Josiah Wedgwood, one year older than Erasmus, founder of the eponymous pottery business. One of the first people to make a fortune out of the Industrial Revolution, Wedgwood married one of his cousins, Sarah, in 1764. She had been born in 1734. The marriage was happy, and decidedly more conventional than the second one of Erasmus, but no less prolific; they produced three daughters and four sons. Their eldest daughter, Susanna, married Erasmus's son, Robert, in 1796, a year after Josiah Wedgwood had died. Sarah survived until 1815.

The marriage, like much of Robert's life, was largely organised by old Erasmus. Eager to have a son succeed him in the medical profession, Erasmus had been heartbroken when Robert's older brother, Charles, died from septicaemia contracted while carrying out a post-mortem as a medical student. The next son, also called Erasmus, was only a year younger than Charles and already committed to a career in law; but Robert, only twelve when Charles died, was young enough to be moulded in the way Erasmus senior wanted, and became a doctor whether he wanted to or not. Pulling strings and providing financial backup, Erasmus saw Robert installed in medical practice near Shrewsbury at the age of twenty, in 1786. Like his father, he was a tall man, tending to corpulence as he grew older. He married when he was thirty and Susanna was a year older than him. Following his marriage, Robert bought land on the outskirts of Shrewsbury and built a new house, The Mount, completed in 1800.

The Darwins' daughters Marianne, Caroline and Susan were born in 1798, 1800 and 1803; their first son, Erasmus, in 1804; Charles Robert Darwin was born on 12 February 1809; and the baby of the family, Emily Catherine (known as Catty or Catherine), arrived in 1810, when Susanna was forty-four. As Darwin's biographer Janet Browne has pointed out, in the early years of the nineteenth century the life of the Darwin family, and the Darwin family itself, could have been lifted straight from the pages of a Jane Austen novel. Already wealthy thanks to his inheritance (and that of Susanna), Robert Darwin made more money through shrewd investments, particularly favouring property and mortgages on land, but soon moving into canals. When he died, in 1848, he left £223,759, an enormous sum in those days (equivalent to several million pounds today).

Josiah Wedgwood's son and heir, Josiah Wedgwood II, bought an estate thirty miles from Shrewsbury, a country retreat called Maer Hall, and the strong links between the Darwins and the Wedgwoods continued into succeeding generations, with several intermarriages.

Josiah II himself married his cousin Elizabeth in 1792, and they had nine children. Their first son, Josiah III, married Charles Darwin's sister Caroline in 1837; the youngest of their brood, Emma, was born in 1808, when Elizabeth was forty-four years old, as Susanna Darwin was when her youngest child was born.

Charles Darwin's idyllic home life, in a large country house, spoiled by older sisters, came to an end in July 1817, when his mother died. Marianne and Caroline were old enough to take over her role in charge of the household, but this left them less time to indulge their younger siblings. Worse, Robert Darwin was so depressed by the loss of his wife that he forbade anyone to mention it, and his gloom pervaded the family.

Charles (who, as a boy, was actually known to his family as 'Bobby') had started school not long before his mother died, having previously been educated at home by his sister Caroline. In 1818, he was sent as a boarder to nearby Shrewsbury School, which he hated. It was only a mile from his home, but he lived in the school, although he got away as often as possible to spend an evening at The Mount. The best thing about the school (although Darwin made many friends among the boys there) was that his older brother, Erasmus (known as 'Eras', or 'Ras'), was a senior pupil, and could keep an eye on him. In spite of the age difference, the two brothers now became close friends, and among other things developed a passion for chemistry, something of a fad in the 1820s.

In 1822, Eras went up to Cambridge to study medicine. Bored by his studies, and a lover of the good life, he became an archetypal Cambridge undergraduate of his day, doing the minimum amount of work and obtaining the maximum amount of pleasure from his time at the university. Charles, who visited him in the summer of 1823, was introduced to this delightful way of life at an impressionable early age. About the same time, still only fourteen, he developed a passion for hunting, and for years (much to his later shame) would take great delight in shooting just about anything that moved. By 1825 his father, Dr Robert, was sufficiently alarmed by the way young Charles was developing that he took him out of school and made him spend the summer assisting in the medical practice, before packing him off to

Edinburgh, at the age of sixteen, to study medicine. This decision by the doctor was both a counsel of despair (he had decided that Charles would never make the grade in an academic subject like law) and the continuation of a family tradition which meant a lot to Dr Robert. But he had more disappointment in store.

To Charles, it was all a great lark, not least because Eras, three years into his own medical studies, had decided to spend his external hospital year in Edinburgh. Very little work got done by either of the Darwins during that year, but Charles did discover that he could never become a doctor. He was physically sick when asked to dissect a corpse, and fled from the room when expected to witness an operation on a child (remember, this was before the days of anaesthetics). When he returned to Edinburgh, without Eras, in 1826, he knew that his medical studies were a sham, and spent much more time in geological pursuits. Geology was a new and exciting science in the 1820s (rather like cosmology in recent times), and Darwin lapped it up. At the end of the academic year, though, in the summer of 1827 he decided that he would have to admit to his father that there was no hope of him ever becoming a doctor. Postponing the evil day, Darwin visited London, where he met up with his cousin Harry Wedgwood, and Paris, where he enjoyed the company of Harry's sisters Fanny and Emma, before heading back to The Mount in August. There were very few options now open to Charles and his father. As Dr Robert saw it, the only hope to save his son from a life of debauchery, squandering his share of the family fortune, was the Church, the traditional safety net for younger sons of gentlemen in those days. After a few months spent cramming Latin and Greek, and reading religious texts, Charles passed (just) the entrance examination for Cambridge, and early in 1828 he was off to begin his studies, towards an ordinary BA degree, the prerequisite for taking Holy Orders. He certainly had no vocation for the Church, but his letters show that he quite liked the idea of being a country clergyman, which would enable him to indulge his interests in natural history and geology and lead a comfortable life (moving on from the time of Jane Austen, we can see many examples of the kind of parson Darwin might have become in the pages of Anthony Trollope's earlier works).

Darwin worked hard enough (at last) to pass his examinations, graduating in 1831 (with a good pass, tenth out of 178), but spent more time studying botany and natural history, under the influence of the Professor of Botany, John Henslow, than on his official courses. This was nothing unusual. In those days, the authorities at Cambridge took no particular interest in exactly what a young gentleman did while he was there (at least, not academically; they took a close interest in the moral side of a young gentleman's life) as long as he passed the exams. Darwin became an expert on beetles, studied geology under Adam Sedgwick, continued to enjoy the hunt, and became firm friends (nobody now can be quite sure how firm) with Fanny Owen, the daughter of one of Dr Robert's neighbours. All good things come to an end, though, and in the summer of 1831 Charles went on one last (as he thought) glorious geological expedition, hammering his way across Wales, before returning to The Mount on the night of 29 August, intending, for lack of any option, to take up the career of a clergyman. Instead, he was just in time to take advantage of an offer that would change his life completely.

The offer was waiting for him in the form of a letter from one of his Cambridge tutors, George Peacock, passing on from Captain Francis Beaufort, at the Admiralty, the invitation to join Captain Robert FitzRoy on a circumnavigation of the globe in HMS *Beagle*. As we have seen, FitzRoy wanted a young gentleman to accompany him on the trip, a civilian social equal to diminish the isolation of command; the obvious role for such an individual, on such a voyage, would be as a naturalist, studying the flora and fauna of the strange lands that they would visit. But, of course, the young gentleman chosen (or his father) would be expected to pay his own way.

It is often suggested that Darwin was chosen for this opportunity not in reality as a naturalist, but simply as an acceptable young gentleman, who liked to hunt. This is far from being the complete truth. To be sure, he was a young gentleman, and he did like to hunt. But he had been recommended by Henslow (who had seriously considered offering himself for the role), who was well aware that Darwin was an able naturalist, had more than a smattering of geology, and would himself benefit from the experience. The one real stroke of luck for Darwin was that Henslow's first choice, another of his protégés, turned the opportunity down because he had just become the vicar of Bottisham. So Henslow wrote to assure Darwin that he felt the young man was 'amply qualified for collecting, observing, & noting any thing worthy to be noted in Natural History . . . I assure you I think you are the very man they are in search of'. Of course, Darwin would never have got the chance if he had not been a personable young gentleman from a wealthy family. But he was also a good naturalist. He was, indeed, the best man for the job.

Darwin's father took a great deal of persuading, and the date set for departure was only a month away. Thinking that he was about to see his problematic younger son settled in a career in the Church, the last thing he wanted was for him to go gallivanting around the world on some harebrained scheme. He only came around to giving reluctant support for the idea (not just moral, but financial support) thanks to the persuasion of his friend Josiah Wedgwood II, who helped Charles to present a clear case to Robert Darwin, answering Robert's doubts and emphasising the benefits of the voyage. Eventually, everything was agreed. Darwin met FitzRoy (who had the final say, of course, in who he wanted to accompany him on the voyage), was kitted out, and joined the *Beagle*. We shall go into the details at the end of this chapter, and the voyage itself is described in the following chapter; but for completeness it seems appropriate to sketch the rest of Darwin's life here – anyone with a passion for chronological continuity can skip to page 115.

After kicking their heels for a few weeks waiting for a fair wind, the *Beagle* sailed on 27 December 1831, heading, initially, for the Canary Islands. Darwin was a few weeks short of his twenty-third birthday. The voyage actually lasted for all but five years, longer than even FitzRoy had anticipated at the outset. While the *Beagle* was busy surveying uncharted waters, Darwin made long expeditions inland, studying not just the living fauna and flora of South America but its geology and fossil remains. He sent back crates and crates of samples to England, wrote long descriptive letters about what he had found to Sedgwick and Henslow, experienced a major earthquake first hand, saw active volcanoes, and saw for himself how geological activity was actively raising the land from the sea. He suffered severe illness, got caught up in revolution, and saw the islands of the South Seas, including the Galápagos. And although he did not know it, by the time the *Beagle* docked in Falmouth, in the night of 2 October 1836, Darwin was already a famous man in scientific circles, his reputation – as a geologist, not any kind of biologist – established by the samples and letters he had sent back from South America, which helped to establish the immense age of the Earth and the nature of the long, slow processes of geological change that have shaped it. He became a Fellow of the Geological Society immediately upon his return to England; but Darwin did not bother about getting elected to the Zoological Society until 1839.

After the social whirl of activity which swept up the returned traveller, he settled down in London to work on two great projects, a book about the geology of South America and his journal account of the voyage (see Chapter Six). In secret, in July 1837 he also started a private notebook on 'the transformation of species'. Around this time, safely settled in London, he wrote to FitzRoy, 'I think it far the *most fortunate circumstance* in my *life* that the chance afforded by your offer of taking a Naturalist fell on me.'[1] He had, of course, no need to worry about anything as mundane as earning a living, and lived off money supplied by his father, who was gratified (if somewhat baffled) to find that his problem son had made something of himself after all (rather more of himself than Eras, in fact, who had abandoned medicine and was living a carefree life in London on an allowance provided by Dr Robert, a clear indication of how Charles might have turned out without the *Beagle* experience).

To his own surprise, Charles turned out to be a gifted writer. The journal of the voyage was not published until May 1839, because it had to wait until FitzRoy finished his share of the book, describing the nautical side of the voyage. To FitzRoy's annoyance, Darwin's account proved so popular that it was split off to form a separate book and reprinted in August, establishing him as a scientist writing for a popular audience about subjects of widespread interest such as the age of the Earth – the Stephen Hawking of his day. He actually made money from his writing, and moved on to a more academic book, about coral reefs. In it, he explained how these atolls are built up by coral growing upwards from the tops of islands that are gradually sinking beneath the waves, an explanation which still stands today. He wrote scientific papers about the new theory of Ice Ages. And all the while, he was thinking deeply, and privately, about evolution. But for the Charles Darwin of the early 1840s, the famous geologist, to have gone public with a theory of evolution would have seemed as bizarre to his colleagues as if some famous cosmologist in the 1980s had come up with a new theory on the origin of life. Before he could give the world the benefit of his insights on the nature of life, Darwin knew that he would have to establish a reputation as a biologist, not just a geologist.

He also had other concerns about going public with these ideas, since he was now a family man, and his wife, Emma Wedgwood, was conventionally religious. He knew that she would be distressed both by his increasingly atheistic views and by the inevitable reaction of the Church to their publication. This was one factor (just about the only one) Darwin had not weighed in the balance when he decided to get married. The idea of marriage, children and domestic comforts appealed to Darwin greatly after the adventure of the voyage of the *Beagle* (indeed, it had begun to appeal to him long before the voyage ended), and in the summer of 1838, in his usual methodical way, he drew up a list of the advantages and disadvantages of marriage. The debit side included the risk of there being 'less money for books &c', while the credit side referred, among other things, to the desirability of a 'constant companion, (& friend in old age) who will feel interested in one, – object *to be* beloved and played with, better than a dog anyhow'. Love was not an essential prerequisite to marriage for Darwin, just as it wasn't important for many of his contemporaries. That could come later. What he wanted was somebody from a suitable background, who he knew and liked, and who preferably brought a bit of money with her. His cousin Emma was, especially given the various intermarriages between Darwins and Wedgwoods, the obvious choice. After an extremely diffident courtship (Emma later wrote that she had 'thought we might go on in the sort of friendship we were in for years') he proposed to her on 11 November, at Maer. To the delight of both families, she accepted at once. She brought with her a dowry of £5,000 plus £400 per year, while Dr Robert bestowed £10,000 upon the happy couple to secure their future. The returned traveller, after five years away, was in no mood to delay the settling down process, and they were married, by yet another cousin, John Wedgwood, on 29 January 1839, at St Peter's Church, Maer – just five days after Darwin had been elected a Fellow of the Royal Society.

While courting Emma, at the end of September 1838 Darwin began reading the *Essay on the Principle of Population*, by Thomas Malthus. The key developments in Darwin's theory of evolution by natural selection seem to have taken place over about the next six months, exactly during the time of his courtship of Emma and the first few months of their marriage. By the summer of 1839, Darwin had most of the theory clear in his head, and much of it recorded, in one way or another, in his notebooks. Malthus was such a key influence that it is worth going into a little detail about just what it was he said in that essay.

The essay was first published, anonymously, in 1798. It dealt with the way populations will increase exponentially if unchecked, because (in human terms) each set of parents is capable of producing more than two offspring – a point strikingly brought home by the fecundity of both the Darwin and the Wedgwood families. At the end of the eighteenth century, the human population of North America was doubling every twenty-five years, a process which clearly could not be sustained for ever. If each pair of even the slowest-breeding land mammals, elephants, left just four offspring that survived and bred in their turn, then in only 750 years each pair would produce 19 million living descendants. In fact, over the 750 years or so leading up to Malthus's work, on average each pair of elephants from 1150 had just two living descendants in 1800. To give an actual example from the human population of the time, if we start with Henry FitzRoy, the first Duke of Grafton, and count his children as generation one, in the fifth generation, which included Robert FitzRoy, Henry had thirty-seven descendants who bore the FitzRoy surname (not all of whom survived infancy) and an untraceable, but probably comparable, number of descendants through the various female lines. Malthus realised that the potential of populations in nature to grow wildly is kept in check by external factors such as predation and the availability of food (and, nowadays, contraception) – not factors which would have checked the growth of aristocratic families in eighteenth-century England. In nature, the *majority* of individuals do not survive to reach maturity and breed. In Darwin's day, this argument was misused in some quarters to claim that the desperate way of life of the working classes in newly industrialised Britain was natural, and that it would be counter-productive to try to eliminate the causes of disease and hunger among the working population, because they would only breed until disease and hunger kept them in check once more.

Darwin saw a broader significance in Malthus's work. He looked at populations in general, not at what was happening in the cities of Britain, and realised that in nature the individuals that did survive and breed would be the ones best adapted to their way of life – best fitted to their ecological niche, like the fit of a key in a lock (not the fittest in the athletic sense, although in some cases that might be one of the relevant factors). The least well adapted would be the losers in the struggle for survival, and would leave no descendants, because they would not live long enough to breed. Provided that there was a variation among the individual members of a species, in each generation the fittest would tend to survive better and leave more offspring. And, crucially, Darwin realised that this struggle for existence is indeed a competition between members of the same species. It is *not* a struggle between, say, rabbits and foxes, but between rabbit and rabbit, or between fox and fox. If one rabbit can run fast enough to escape the fox, but a second rabbit cannot, it is the fast runner who survives (in this case, physical fitness does matter!). The fox doesn't mind which rabbit it eats, as long as it gets food.

By the end of 1838, even before he was married, Darwin's notebooks show that he was already drawing the comparison between this process of natural selection and the artificial selection process used by plant and animal breeders to 'improve' horses, or greyhounds, or wheat. And early in 1839 he was spelling out the realisation that there is no need for nature to 'know' which variations on the theme are good ones in order for it to work. If it just happens that a bird is born with a slightly longer beak than average, and that variation helps it to get more food, then it is likely that the bird will survive and breed, passing on its tendency for a slightly longer beak to the next generation. Blind chance, operating on individuals, combined with a tendency for characteristics to be inherited (but imperfectly, so that there is always the chance for new variations to arise) is all you need to explain evolution.

This was Darwin's great original idea – not evolution, which was already a well-known (if not always accepted) idea, but *natural selection*, what became known as the 'survival of the fittest'. Critics of Darwinian ideas sometimes refer to 'the theory of evolution', but they are wrong to do so. Evolution is a fact. The full name of Darwin's theory is 'the theory of evolution by natural selection', so if you do want to use a shortened version of the name, you should refer to the theory of natural selection, which you may or may not want to criticise. Darwin had yet to give the theory its name, but the clarity with which he had understood what was going on by the beginning of the 1840s is made clear from the notebooks.

And yet, he sat on this idea for nearly twenty years, hugging it to himself and keeping it from the world. This didn't stop him working on it. A thirteen-page outline of the theory, undated but seemingly from 1839, survives in his papers, together with what he later referred to as a 'brief abstract', running to thirty-five pages, dated 'May & June 1842'. In the spring of 1844, he developed this version of the theory into a description about 50,000 words long, and had a fair copy of it made; he wrote a letter to Emma, to be opened in the event of his death, asking her to have this version published. But he seems at that time to have had no intention of publishing in his lifetime.

And yet . . . Darwin, although not concerned about fame in his lifetime, did have the scientist's urge for his priority to be recognised by posterity, when and if the theory was discovered by someone else. In order to ensure this, he devised a plan so cunning that it was more than a hundred years before it was uncovered. During 1845, Darwin worked on a second edition of his successful journal of the *Beagle* voyage. In this, he added bits of new material to the existing descriptions of the living things he had seen in South America. All of these new passages look innocuous in their places in the book. But as Howard Gruber pointed out in his book *Darwin on Man*,[2] if you compare the first and second editions of the journal, you can locate all the new material, take it out of the second edition and string it together to make a coherent 'ghost essay' which conveys almost all of Darwin's thinking about evolution at the time. It is quite clear that this material must have been written as that coherent essay, then carefully chopped up and inserted into the journal. Darwin, it seems, was torn between keeping quiet about his theory and telling the world. The situation was complicated by his personal life – first, as we have seen, by his desire to avoid hurting Emma. Also, Darwin himself had been plagued by illness, off and on, ever since he returned from his travels, and he would be for the rest of his life. Whole books have been written about this illness, some suggesting that it was entirely psychosomatic, brought on by worrying about the likely public reaction to his ideas about evolution, others seeking a physical cause of one kind or another. The most likely explanation seems to us to be that it was a genuine physical ailment, very probably something he had picked up in the tropics, which was exacerbated by overwork and worry. Either way, Darwin did not see himself as the kind of robust individual that would be needed to champion these ideas publicly.

The family, too, was growing, and he had an increasing number of dependants. Charles and Emma produced ten children between 1839 and 1856 (when Emma was forty-eight), seven of whom survived into adulthood (one, Leonard, lived until 1943). Two died in infancy; one, Darwin's favourite daughter Annie, died in 1850, at the age of ten, plunging Darwin into a deep depression.

There was also unrest in the outside world in the 1840s and 1850s, sufficient on its own to make Darwin hesitate about poking his head above the parapet. Although he had settled down with Emma in London after their marriage, by 1840 there were riots in the streets as the Chartist movement sought reform of the electoral system, and the Darwins decided that London was no place to bring up their family. With financial assistance from Darwin's father, they purchased Down House, in the village of Down, in Kent, and moved there in the middle of September 1842, a month after the army had been out on the streets of London subduing the latest violent demonstrations. A few years after the move, which made Darwin a member of the squirarchy, the spelling of the village name changed to Downe, but the name of Down House remained the same; Darwin lived there for the rest of his life, only two hours by carriage from the centre of London, but deep in the heart of the Kentish countryside. In spite of the growth of roads and spread of houses since then, it still remains, almost as he left it.

A week after the move, Emma gave birth to their third child, Mary Eleanor, who survived for only three weeks. It was a gloomy start to what turned out to be a long and happy life in Down House; but William (born in 1839) and Annie (born in 1841) thrived in the country environment and, almost as if to prove the Malthusian point, within three months of losing Mary Eleanor, Emma was pregnant again. So it was in Down House that Darwin, surrounded by a growing family and with all the domestic comforts he had craved, completed, for the benefit of posterity, his 1844 summary of the theory of evolution by natural selection and started making his name as a biologist, to make sure his idea was taken seriously. He had written a book about volcanic islands, which was published in 1844, and his long-delayed *Geological Observations on South America* was published in 1846, ten years after he had returned to England in the *Beagle*. In October that year, he wrote to Henslow:

> You cannot think how delighted I feel at having finished all my Beagle materials except some invertebrata; it is now ten years since my return, and your words, which I thought preposterous, are come true, that it would take twice the number of years to describe, than it took to collect and observe.

The invertebrata that Darwin referred to in that letter were some peculiar barnacles that he had picked up on the shores of southern Chile in 1835. Thinking that they would make a fine study to establish his credentials in biology, and intrigued in any case by these bizarre creatures, each the size of a pinhead, that lived within the shells of other barnacles, he set to work enthusiastically. But, in order to explain the place of these creatures in the barnacle family, it turned out that he had to work out the complete classification of barnacles, which was a hopeless mess. It was five years before the first of his epic volumes on barnacles appeared in 1851, produced in spite of bouts of crippling illness, Annie's death, and Darwin's increasing despair at the Herculean task he had taken on. In 1854, the final two volumes of what had turned out to be a trilogy appeared. Together, they still form the definitive work of barnacle taxonomy, and they were received with acclaim by the scientific community, earning Darwin (together with his work on coral reefs) the Royal Medal of the Royal Society. It had been far harder work than he had imagined, and taken far longer than he had expected; but by the end of the 1850s Charles Darwin was established as a biologist of the first rank, one who understood all about the relationships between species.

In the mid-1850s, Darwin still had no intention of publishing his theory of natural selection. He was forty-six in 1855, a middle-aged semi-invalid with responsibilities that extended beyond his own family. His father had died on 13 November 1848, and in the months that followed Darwin's own illness became so severe that he resorted to the fashionable 'water cure', involving early morning sessions in which he was first wrapped in blankets and heated with a spirit lamp until he was running with sweat, then plunged into ice-cold water. He also went on a special diet, without any sugar, salt, bacon or alcohol, and was allowed to work for only a few hours a day – one reason why the barnacle study took so long. Whatever the reasons, the treatment seemed to work, and Darwin stuck with it, more or less, for most of the rest of his life, although he still suffered bouts of chronic illness.

In the outside world, 1851 saw the Great Exhibition in the Crystal Palace. On 13 May, only three weeks after Annie had died, Emma gave birth to the Darwins' ninth child, Horace; so it was in July that the entire family spent a week staying with Erasmus in London and visiting the Exhibition. The visit also gave Darwin an opportunity to renew his friendship with Joseph Hooker, one of the few people he had confided his theory of natural selection to. Hooker was a young botanist (he had been born in 1817), the son of the Director of the Royal Botanic Gardens at Kew, who Darwin had met at the end of 1843, shortly after Hooker had returned from an expedition to the Antarctic, as naturalist with a naval expedition. He became one of Darwin's closest scientific confidants, someone ideas could be bounced off; it had been a great blow to Darwin when Hooker went off on another expedition, to the Himalayas, in 1847, and Darwin was delighted to have him back.

By now the Darwins were extremely wealthy (Charles had inherited £40,000 from his father to add to their steadily accumulating fortune) with investments in industry, railways and land. But the need to manage these investments was a constant source of worry to Darwin, who also sought to be a model landlord, looking after his tenant farmers, and diligently carried out his duties as a gentleman around Downe, helping the poor and needy. These worries were exacerbated by the Crimean War, which started in 1854 and which led to the fall of the government and economic uncertainty in Britain. On top of all this, in December 1856, at the age of forty-eight, Emma gave birth to their last child, Charles Waring – and he turned out to be suffering from what seems with hindsight to have been Down's syndrome. The baby would die less than two years later, in June 1858, of scarlet fever. And that death would come when Darwin was in the midst of a frantic burst of work triggered by the discovery that while he had been prevaricating and hiding ghost essays among his papers for later generations to discover, someone else had come up with the idea of natural selection.

Darwin had not been completely scientifically inactive in the mid-1850s. For public consumption, he had carried out work on how seeds could spread around the world, testing how long they could survive in salt water and still sprout, the way in which they could spread through bird droppings, and even floating a dead pigeon in salt water for thirty days before recovering seeds from its crop and growing plants from them. These were important studies, which nobody before had bothered to carry out, and emphasise the way in which Darwin was a 'hands-on' naturalist, not an abstract armchair thinker. When he wanted to know more about artificial selection, for example, he joined pigeon-breeding societies, and learned first hand about the way in which characteristics are passed on from one generation to the next, and how different variations on the same basic body plan can be selected by the breeder. He really did understand how living things behaved, from direct, practical experience.

Privately, at this time Darwin was working on a book-length version of his theory of evolution by natural selection. He was coming round to the view that he might publish this eventually, when he was so old that it could do him no harm, and mentioned his progress with the epic work from time to time to the few people in the know. As well as Hooker, these included the geologist Charles Lyell, who had been a major influence on Darwin at the time of the voyage on the *Beagle* (Lyell, born in 1797, wrote an epic three-volume book of his own, the *Principles of Geology*: Darwin took volume one with him on the *Beagle*; volume two caught up with him during the voyage; volume three was waiting for him when he returned home). Another of Darwin's circle at this time was Thomas Henry Huxley, who he had met at a meeting of the Geological Society in April 1853. Another naturalist who had cut his teeth on a long voyage with a naval surveying ship, Huxley was an impoverished young firebrand (he had been born in 1825) who raged at the way science in England seemed to be the preserve of gentlemen of independent means (like Darwin), but was shrewd enough to appreciate the need for patronage from the likes of Darwin, and quickly came to realise that here was a kindred revolutionary spirit, at least in scientific terms, for all his wealthy background. Huxley and Hooker, by the mid-1850s, were also close friends, members of a young generation that would indeed make science a real profession, instead of the plaything of gentleman amateurs.

Darwin's own state of mind about his evolutionary ideas can be gleaned from a letter he wrote to Lyell in November 1856: 'I am working very steadily at my big book; I have found it quite impossible to publish any preliminary essay or sketch; but am doing my work as completely as present materials allow without waiting to perfect them.' The key word is 'completely'. Darwin wanted to dot every *i* and cross every *t* before considering publishing his ideas. He wanted to consider every possible argument against his theory, and find a counter-argument that could appear in his epic book, leaving would-be critics without a leg to stand on. He was not going to go off half-cock, publishing a version of the theory which would be correct in broad outline, but might contain errors of detail with which opponents could damn him and the theory.

If Darwin had had his way, the theory of natural selection would eventually have been published, perhaps not until after his death, in a massive three-volume tome, densely packed with examples and arguments, and destined to be read only by a few scientists. The fact that he was forced by circumstances into rushing into print what he considered to be merely an outline of the theory, an abstract written in beautifully clear, accessible language, a book which became an instant best-seller and has remained in print ever since, is one of the greatest strokes of fortune there has ever been for science. It happened like this. Alfred Russel Wallace was a naturalist who, in 1858, was based in the Far East. He had been born in 1823, the eighth of nine children, and had had to leave school at thirteen to earn a living. After an eventful early life, including work as a surveyor on the canals and as a schoolteacher, Wallace became a kind of freelance naturalist, supporting himself by selling specimens that he sent back to England to just those wealthy gentleman amateurs that Huxley (in principle, although, as we have seen, not always in practice) despised (including Darwin). Wallace came to the theory of natural selection by exactly the same route that Darwin had, through experiencing the fertile workings of nature in the tropics, and even through combining these direct observations with a reading of Malthus's *Essay*. Wallace corresponded with Darwin, initially mainly about specimens, not realising that Darwin had already come up with the theory of natural selection. In September 1855, Wallace published a paper in the journal *Annals and Magazine of Natural History*, which presented some of the evidence for evolution, but without introducing the idea of natural selection.

Some of Darwin's friends were sufficiently alarmed by the similarities of parts of this paper to aspects of Darwin's own work that they urged him to publish something himself. Lyell, who was by no means convinced that Darwin was right, but had the true scientist's instinct for priority, wrote to him on 1 May 1856, saying: 'I wish you would publish some small fragment of your data, *pigeons* if you please and so out with the theory and let it take date and be cited and understood.' Darwin, though, wasn't worried about Wallace, who he felt, from reading Wallace's paper, had some good ideas but still favoured the view that species were created by God. Evolution, after all, was nothing new; and there was no hint in Wallace's paper of natural selection. Although, under pressure from his friends, he began to consider writing a shorter account of his own work, for publication, as late as May 1857 he was writing to Wallace in sublime ignorance of how far Wallace had gone, sending him a carefully worded 'keep off' notice, intended to establish that Darwin was the leader in these matters. Although he handed Wallace the sop of commenting 'we have thought much alike and to a certain extent have come to similar conclusions', he went on:

> This summer will mark the 20th year (!) since I opened my first note-book, on the question how and in what way do species and varieties differ from each other. I am now preparing my work for publication, but I find the subject so very large, that though I have written many chapters, I do not suppose I shall go to press for two years.

Wallace didn't take the hint. If anything, what he took as Darwin's endorsement of his ideas encouraged him to write up a full account of them, including natural selection (although he did not give it that name). The result, a paper titled *On the Tendency of Varieties to Depart Indefinitely from the Original Type*, was mailed to Darwin with a covering letter asking him to show it to Lyell. It landed on Darwin's desk on 18 June 1858, prompting him to send the manuscript on to Lyell with a covering letter of his own: 'Your words have come true with a vengeance – that I should be forestalled . . . I never saw a more striking coincidence; if Wallace had my MS sketch written out in 1842, he could not have made a better short abstract!'

Of course, it was not a coincidence; it was a scientific truth, waiting to be discovered by anyone who had eyes to see. Darwin's immediate reaction was to give up any hope of establishing his priority, although he did hope that his book might still be useful as an explanation of the application of the theory. His friends disagreed. Lyell and Hooker, scheming independently of Darwin (now distracted by the illness and then, on 28 June, death of baby Charles), came up with a plan to present a 'joint paper' by Darwin and Wallace to the next meeting of the Linnean Society, due on 1 July. Darwin's contribution was largely based on his unpublished 1844 essay (Darwin had, incidentally, been thirty-five in 1844, the same age Wallace was in 1858); Wallace's contribution was essentially the paper he had sent to Darwin. Wallace, on the other side of the world, wasn't even consulted, but always accepted the fait accompli with good grace, counting himself lucky to be mentioned in the same breath as Darwin.

But after all the fuss in Darwin's inner circle, the joint paper went down like a lead balloon. Almost a year later, at a meeting held on 24 May 1859, the President of the Linnean Society summed up the past twelve months of the society's activities with the words: 'The year which has passed . . . has not, indeed, been marked by any of those striking discoveries which at once revolutionise, so to speak, the department of science on which they bear.' But before the end of 1859, the cat was well and truly out of the bag. Urged on by Lyell, and worried that Wallace might publish a book, Darwin dropped everything in the summer of 1858 and wrote his masterpiece, lifting large chunks from the existing chapters mentioned in his letter to Wallace, revising and honing his 'abstract'. The way his mind was running, and his plans to publish a full version of the theory later, can be seen from the original title, *An Abstract of an Essay on the Origin of Species and Varieties through Natural Selection*. His publisher, John Murray, managed to persuade Darwin to reduce this to *On the Origin of Species*, but the author insisted on keeping *by Means of Natural Selection* as a kind of subtitle, and including on the title page the words *or the Preservation of Favoured Races in the Struggle for Life*. Given his head, Darwin would have put the whole theory in the title! As for being an abstract, the first edition of the book, published in November 1859, ran to more than 150,000 words.

Murray did not expect to have a best-seller on his hands. He initially planned to print only 500 copies of what he thought would be a scientific tome, but increased this to 1,250 when he read the finished manuscript. They were all in the shops on the day of publication, and a new edition followed immediately. Darwin continued to make changes as the book went through several editions in his lifetime. Most of the changes were an attempt to shore up the theory against criticisms which are now known to be misguided. In particular, Darwin never knew how heredity works (DNA was not identified as the carrier of the genetic code until the 1950s) and went further and further up a blind alley trying to explain this. The result is that the definitive edition of the *Origin*, as it is always referred to, is indeed the first, and that is the one you should seek out if you want to learn about natural selection and evolution from the horse's mouth – but don't look for it in antiquarian bookshops; it has been reprinted many times, and is available in a cheap Penguin edition. The theory contains two key ingredients. Individuals in one generation reproduce to make new individuals in the next generation that are similar, but not identical, to their parents. That provides variety. Then, natural selection ensures that the individuals best fitted to their environment reproduce most effectively in their turn and leave more offspring. That's all there is to it. But the implication is that all life on Earth, including human life, has arisen in the same way from some common ancestor. So people are just one species among many, not specially chosen or created by God. And this was where FitzRoy would, as we shall see later, take exception to his former messmate's ideas.

Life would never be quite the same for Darwin after the publication of the *Origin*. By and large, he kept his head down in Kent and left the promotion of the idea of evolution by natural selection to his supporters such as Hooker and Huxley – ironically, Huxley's fervent enthusiasm for the idea (which led to him being described as 'Darwin's bulldog') meant that he was promoting the work of an independently wealthy gentleman scientist, to some extent at the expense of a working professional scientist, Wallace. There is no need to go into details of the debate that raged around Darwin's (and Wallace's) theory in the 1860s; history records who won that debate and why – because the facts were, and are, on the side of natural selection.

With the great work off his chest at last, and in the safe hands of a younger generation of disciples, Darwin continued more or less where he had left off after the rude interruption caused by the arrival of Wallace's paper in 1858. He had long been fascinated by orchids, and the way in which they are fertilised by insects that are superbly fitted by evolution for that task; this resulted in a book, bearing the archetypal Darwin title *On the Various Contrivances by which British and Foreign Orchids are Fertilised by Insects, and on the Good Effects of Intercrossing*, published in 1862. Over the next few years he prepared another magnum opus, *The Variation of Animals and Plants under Domestication*, which was published early in 1868, a few weeks before Darwin's fifty-ninth birthday, and spelled out the details of artificial selection, and thereby, by analogy, provided more ammunition for the arguments in support of natural selection. He began to receive honours from abroad during the 1860s (although, curiously, he was never properly honoured in England during his lifetime, except in strictly scientific circles like the Royal Society).

At home, everything had settled down nicely. The surviving children were all well, and the boys all went on to successful careers in professions as diverse as banking, the Army and, in the case of Francis and George (who were both, unlike their father, knighted in recognition of their achievements) science. There were blacker moments, including FitzRoy's estrangement from Darwin, following the publication of the *Origin*, the deaths of Darwin's sister Catherine in 1866 and Susan a few months later. Professionally, though, a clear sign of which way the wind was blowing came in the summer of 1868, when Hooker was elected President of the British Association for the Advancement of Science. As Darwin entered his own sixties, he might have been expected to rest on his laurels in easy retirement, enjoying what family life was left to him. Instead, he at last produced his second famous book, the long-awaited volume on humankind's (or, as he put it, 'man's') place in evolution, which he had promised in the sentence in the *Origin* in which he said 'light will be thrown on the origin of man and his history'.

Incidentally, if you are worried about his use of the term 'man' for humankind, which today might be regarded as reactionary and sexist, ponder this – Darwin was actually a far-seeing revolutionary, not a reactionary, simply by using the lower-case 'man' instead of the capitalised 'Man' in the text of his book, deliberately ignoring the tradition, in those days, of using the capital M to denote the importance of our species. Fashions change, and what matters is to see these terms in the context of their times.

The light that Darwin himself shed on the subject of the origin of man came in 1871, in a book that was really, as its title implies, two books in one – *The Descent of Man, and Selection in Relation to Sex*. Like the *Origin*, it is usually referred to by a single word, as the *Descent*. In spite of Darwin's best efforts as a skilful writer to weave the two halves together in chapters that describe sexual selection at work in people, *Selection in Relation to Sex* really follows as an afterthought to *Variation*, and *The Descent of Man* stands alone as a summary of humankind's place in evolution. Again, this is not the place to labour the point of what is now, by and large, a non-controversial theory. But the importance of Darwin's approach to the subject, in the context of his time, can best be put in perspective from the words in an essay he actually wrote in 1839, but did not publish. He said (and note the use of the capital M in this earlier work): 'Looking at Man, as a Naturalist would at any other Mammiferous animal, it may be concluded that he has parental, conjugal and social instincts, and perhaps others.'

This is the nub of what Darwin did that was new to science – *looking at Man, as a Naturalist would at any other Mammiferous animal*. No special pleading, no role for God, no suggestion that humankind is in any moral way 'superior' to other animals. With those words, Darwin became the first sociobiologist. And that is the approach which carries through into the *Descent*, thirty-two years later, and which, combined with Darwin's usual clear prose, made the book another best-seller:

> Unless we wilfully close our eyes, we may, with our present knowledge, approximately recognise our parentage; nor need we feel ashamed of it. The most humble organism is something higher than the inorganic dust under our feet; and no one with an unbiased mind can study any living creature, however humble, without being struck with enthusiasm at its marvellous structure and properties.

This was Darwin's last great book, and perhaps it is just as well that FitzRoy did not live to read it. But Darwin's enthusiasm for the marvellous structure and properties of the humblest living organism shines through in a torrent of books that now poured from the pen of the master – a monograph on climbing plants, published in 1875; *Insectivorous Plants* (1875); *The Effects of Cross and Self-Fertilisation in the Vegetable Kingdom* (1876); *The Different Forms of Flowers on Plants of the Same Species* (1877); *The Power of Movement in Plants* (1880); and

The Formation of Vegetable Mould, through the Action of Worms, with Observations on their Habits, published in 1881, the year before he died at the age of seventy-three. He also published five scientific papers in the last year of his life. The biggest sensation of all Darwin's writings in the last decade of his life, though, had been *The Expression of Emotions in Man and Animals*, which sold 5,000 copies on the day of publication, in 1872. What looks like an addendum to the *Descent* found a ready market with Victorians intrigued by the notion that people really did react to outside stimuli in the same way as other animals, and no doubt titillated by the racy chapter on blushing, which included the following description of the 'expression of emotion' in a female patient of one Dr Browne. The lady was in bed when Dr Browne and his assistants visited her, and:

> The moment that he approached, she blushed over her cheeks and temples; and the blush spread quickly to her ears. She was much agitated and tremulous. He unfastened the collar of her chemise in order to examine the state of her lungs; and then a brilliant blush rushed over her chest, in an arched line over the upper third of each breast, and extended downwards between the breasts, nearly to the ensiform cartilage of the sternum.

But don't run away with the idea that this description represents prurience on Darwin's part. His ability to act as a detached scientific observer had long extended even to his own most intimate emotions and instincts, and while courting Emma back in 1838 he had made notes in his notebooks about his reactions, commenting on the way 'sexual desire makes saliva to flow' and the way blushing seemed to have sexual connotations, jotting down, almost in shorthand, 'blood to surface exposed, face of man . . . bosom in woman: like erection'.

In his declining years, Darwin settled into a steady routine at Down House, rising early and taking a walk before breakfast. He worked for an hour and a half, between 8 and 9.30, then read his letters, or rather had them read to him as he lay on a sofa, before going back to work for an hour or so and then taking another walk. After lunch, he read the newspaper, then it was time for him to write his letters, using a board to support the paper as he sat in a chair by the fireplace. Another hour's work in the late afternoon, perhaps reading a novel, dinner, and two games of backgammon with Emma, before bed at half-past ten.

The settled routine was described by Francis Darwin in his *Reminiscences*, which he added to the autobiography of his father that he edited, and which was first published in 1887. Francis had become very close to his parents in those years, because his wife, Amy, died in 1876, soon after giving birth to Charles and Emma Darwin's first grandchild, Bernard. He came back to live in Down House, and acted as his father's assistant until Charles died.

There were family holidays, and an idyllic, sunny summer at Downe in 1881, recalled with affection by the younger Darwins years later. But Charles Darwin was beginning to outlive many of his friends and relations, and on 26 August that year he learned that Eras had died. The news plunged Darwin into yet another bout of illness, in spite of which he made the effort to sort out his brother's affairs, arrange the funeral, and look after the requirements of the will. The effort exhausted him, and he never really recovered, being bedridden for much of his last winter. He died, in Emma's arms, on 19 April 1882. Better, indeed, than a dog.

Although the family intended that Charles Darwin should be buried quietly in Downe, the establishment which had neglected to offer him even a knighthood in life decided (partly thanks to lobbying by Huxley) that it was time to make amends. With the permission of Emma and William (now the head of the family), Darwin was buried in Westminster Abbey, on 26 April 1882. But his life, and death, would have been very different if he had not, as a young man of twenty-two, set sail in the *Beagle* with Robert FitzRoy, at the end of 1831.

The invitation that Darwin received on his return home on 29 August 1831 had arrived thanks to a close-knit network of scientifically inclined men. The George Peacock with whom Beaufort discussed FitzRoy's proposal was a Cambridge mathematician, a Fellow of Trinity College, who was living in London during the University vacation. Beaufort, newly established in his post as Hydrographer, was in the early stages of his successful efforts to modernise the Navy and give it, in the form of surveying, a proper peace-time role. His circle included not only Peacock, but people like the astronomer John Herschel (son of William Herschel, who, with his sister Caroline, discovered the planet Uranus), and Charles Babbage, the computer pioneer. They were all involved in much needed reform of the Royal Society in particular, and the way science was done in Britain in general. In 1831, the planned *Beagle* voyage was by far Beaufort's biggest project to date, and he threw himself behind it 100 per cent, enthusiastically supporting the idea of making it a scientific, as well as a hydrographic, expedition. He conveyed that enthusiasm to Peacock (perhaps a little too enthusiastically, since at first Peacock thought that Beaufort was seeking an official, Admiralty-paid naturalist; the misunderstanding was soon rectified), who wrote to Henslow. Had he been ten years younger, Henslow would have leapt at the opportunity himself. But he was now thirty-five, married, and had a new baby. Reluctantly, he passed the invitation on to Leonard Jenyns, a well-thought-of young naturalist, a Cambridge graduate and a friendly rival of Darwin in the collecting game during Darwin's undergraduate years, who, as we have noted, turned it down with equal reluctance since he had only recently taken up his clerical duties in Bottisham. It was only then that Henslow conveyed the invitation to Darwin – by now with some urgency, since the *Beagle* was due to sail in a month's time.

Henslow's letter to Darwin, dated 24 August, took his acceptance for granted:

> . . . I shall hope to see you shortly fully expecting that you will eagerly catch at the offer which is likely to be made to you of a trip to Terra del Fuego & home by the East Indies – I have been asked by Peacock who will read & forward this to you from London to recommend him a naturalist as companion to Capt Fitzroy employed by Government to survey the S. extremity of America. I have stated that I consider you to be the best qualified person I know of who is likely to undertake such a situation – I state this not on the supposition of y^r being a *finished* Naturalist, but as amply qualified for collecting, observing, & noting anything worthy to be noted in Natural History. Peacock has the appointment at his

disposal & if he can not find a man willing to take the office, the opportunity will probably be lost. Capt. F wants a man (I understand) more as a companion than a mere collector & would not take any one however good a Naturalist who was not recommended to him likewise as a *gentleman*.

Darwin was as eager to take up the offer as Henslow knew he would be, but his hopes were dashed by his father's refusal to grant permission, and he wrote sorrowfully to Henslow explaining the situation. That refusal was, however, tempered by the remark 'if you can find any man of common sense who advises you to go, I will give my consent'. Robert Darwin was referring, as Charles knew, to Josiah Wedgwood.[3] On 30 August, Charles set off to visit his uncle Jos, carrying a letter from Robert Darwin which said:

Charles will tell you of the offer he has had made to him of going for a voyage of discovery for 2 years. – I strongly object to it on various grounds, but I will not detail my reasons that he may have your unbiassed opinion on the subject, & if you feel differently from me I shall wish him to follow your advice.

As history records, Josiah Wedgwood did feel differently, and Robert Darwin did then give his wholehearted support to the project (not just his consent, but considerable financial support). Within a couple of days, outrunning the postal service, Charles was in Cambridge, scribbling a note to Henslow from the Red Lion inn:

I am just arrived: you will guess the reason, my Father has changed his mind. – I trust the place is not given away. – I am very much fatigued & am going to bed. – I daresay you have not yet got my second letter. – How soon shall I come to you in the morning. Send a verbal answer. Good night.

Now, the network sprang into action in reverse, sending news about Darwin back to FitzRoy. Beaufort immediately recognised the name. Erasmus Darwin, Charles's grandfather, had had a close friend called

Richard Lovell Edgeworth, who shared Erasmus's appetites and had four marriages producing a total of twenty-two children. Born in 1743 (making him twelve years younger than Erasmus), Edgeworth entered into the last of these marriages in 1798, when his bride was Miss Frances (Fanny) Beaufort – the twenty-nine-year-old sister of Francis Beaufort. Edgeworth and Beaufort became firm friends, and it was Edgeworth who provided Beaufort's entrée into the scientific world (indeed, it was Edgeworth who was his proposer when Beaufort was elected as a Fellow of the Royal Society in 1814). In addition, although Beaufort had never met Charles Darwin, thanks to this connection in 1803 he had brought his own mother to England from Ireland to consult Dr Robert Darwin about a skin complaint. So it was that as early as 1 September 1831 (clearly anticipating Darwin's acceptance of the offer without having yet heard from him) Beaufort would write to FitzRoy, 'I believe my friend Mr Peacock of Trin^y College Camb has succeeded in getting a "Savant" for you – a Mr Darwin, grandson of the well known philosopher and poet – full of zeal and enterprize.'

About this time, however, FitzRoy seems to have had second thoughts about living on intimate terms with a complete stranger for several years. He said that he had already asked a friend, named Chester, to accompany him, and that there would be no room on board for Darwin after all. FitzRoy did have a friend called Harry Chester, a son of Sir Robert Chester, who was a clerk in the Privy Council Office at the time. But it is not at all clear whether he ever was offered the chance to go on the voyage (let alone whether he accepted such an offer), and it may well be that FitzRoy was only looking for a polite way of rejecting Darwin (or anyone else put forward by the Cambridge network) if he didn't like the cut of his jib. This suspicion is strengthened by the fact that he did agree to meet Darwin, and that once it became clear that they got on like a house on fire, 'Mr Chester' disappeared like the exploded Bunbury in *The Importance of Being Earnest*. During their first meeting, after sizing Darwin up FitzRoy announced that just five minutes earlier he had received a note saying that Chester was 'in office' and could not go after all.

Darwin's immediate reaction to FitzRoy amounted to hero worship. He wrote immediately to Henslow:

Gloria in excelsis, is the most moderate beginning I can think of.
– Things are more prosperous than I should have thought possible.
– Cap. Fitzroy is everything that is delightful, if I was to praise
half so much as I feel inclined, you would say it was absurd only
once seeing him. – I think he really wishes to have me. – He offers
me to mess with him & he will take care I have such room as is
possible . . . Ship sails 10th of October . . . What has induced Cap.
Fitzroy to take a better view of the case is; that Mr. Chester, who
was going as a friend, cannot go: so that I shall have his place in
every respect.

To his sisters, he often referred to FitzRoy as his 'Beau Ideal' of a
Captain, and clearly by no means entirely in jest. His granddaughter,
Nora Barlow, has summed up the influential nature of the relationship
from a Darwin point of view:

Though the circumstances of the voyage must be reckoned as
the main factor of those decisive years, yet I think it can be
maintained that the personal element counted also, and that the
close companionship of a man of FitzRoy's moral ascendancy
and intellectual intransigence left their indelible mark. Darwin
was not yet twenty-three when he set sail, on the threshold of
his career, whilst FitzRoy was only twenty-six, years when
personal influences count for much, and Darwin from the very
beginning fell under his senior's charm. FitzRoy possessed an
integrity of outlook that made an instant appeal to Darwin.[4]

FitzRoy himself wrote to Beaufort after his first meeting with Darwin:

I like what I see and hear of him, much, and I now request that
you will apply for him to accompany me as a Naturalist. I can
and will make him comfortable on board, more so perhaps than
you or he would expect, and I will contrive to stow away his
goods and chattels of all kinds and give him a place for a
workshop.

In order to obtain the necessary goods and chattels, Darwin indulged in a frantic burst of shopping, including trips to London; but everything had to stop for one special occasion. As he wrote to his sister Susan on 9 September:

Yesterday all the shops were shut [for the Coronation of William IV], so that I could do nothing – and I was child enough to give £1. 1s. for an excellent seat to see the procession – and it certainly was very well worth seeing. – I was surprised that any quantity of gold could make a long row of people quite glitter. – It was like only what one sees in picture-books of Eastern processions.

When the shopping did resume, it is truly astonishing what FitzRoy and his officers did indeed manage to stow away in the *Beagle*. Darwin's share of the space was a corner of the tiny poop cabin (the whole cabin was about 10 feet by 11 feet, and too low to stand up in), also occupied by John Lort Stokes and Philip Gidley King. The space was dominated by a chart table, but each had his own corner to sit and read or work. At night, Stokes slept in a bunk under the steps at the entrance to the cabin, while Darwin and King slung their hammocks across the chart table. To give you some idea of how modest the space available was, in order to fit his lanky frame in, each night Darwin had to remove one of the drawers that lined the wall of the cabin to make room for the end of the hammock containing his feet. As if that were not enough, the cabin doubled as the ship's library, storing all the books brought on board by the officers and Darwin. One estimate[5] is that there were nearly 250 volumes (some of them, admittedly, no more than pamphlets, but also including the *Encyclopaedia Britannica*), all of which could be borrowed by the officers, under the supervision of the Captain's clerk.

Darwin's regular escape from these cramped conditions when on board (but remember that more than half of his time during the voyage was actually spent ashore) was to take his meals with the Captain. On 13 December, even before the (much delayed!) ship had left harbour, he would write in his diary: 'Dined for the first time in Captains cabin & felt quite at home. – Of all the luxuries the Captain has given me, none will be so essential as that of having my meals with him.' Much later, in a letter home he describes a typical day at sea:

> We breakfast at 8 o'clock. The invariable maxim is to throw away all politeness: – that is, never to wait for each other, and bolt off the minute one has done eating, etc. At sea, when the weather is calm, I work at marine animals, with which the whole ocean abounds; if there is any sea up, I am either sick or contrive to read some voyage or travels. At one we dine. You shore-going people are lamentably mistaken about the manner of living on board. We have never yet (nor shall we) dined off salt meat. Rice and Pea and Calavanses [a variety of pulse] are excellent vegetables, and with good bread – who could want more? Judge Alderson[6] could not be more temperate, as nothing but water comes on the table. At 5 we have tea.[7]

But Darwin did get to see another side of FitzRoy's character before the voyage began. In his autobiography, he writes:

> At Plymouth before we sailed, he was extremely angry with a dealer in crockery who refused to exchange some article purchased in his shop: the Captain asked the man the price of a very expensive set of china and said 'I should have purchased this if you had not been so disobliging.' As I knew that the cabin was amply stocked with crockery, I doubted whether he had any such intention; and I must have shown my doubts in my face, for I said not a word. After leaving the shop he looked at me, saying You do not believe what I have said, and I was forced to own that it was so. He was silent for a few minutes and then said You are right, and I acted wrongly in my anger at the blackguard.

What is significant is that in recounting this anecdote many years later, Darwin is holding it up as an example of FitzRoy's candour and willingness to acknowledge his own faults.

It's hardly surprising that the preparations for the voyage took longer than expected, and even after FitzRoy received his official sailing orders from the Admiralty on 15 November, bad weather kept them in harbour for weeks. At least this did give Darwin and FitzRoy a chance to get to know one another better, and to confirm that they got on well enough (in spite of incidents like the one in the china shop) for Darwin to take his place on board the *Beagle*. There would be one, less than satisfactory, Christmas to endure in Plymouth before the ship at last got free from the harbour as 1831 was drawing to a close.

CHAPTER FIVE

The Darwin Voyage

A FTER ALL THE delays, the westerly gales abated and *Beagle* was finally able to set out from Plymouth on her epic voyage on 27 December 1831. One side effect of the enforced stay in harbour over Christmas immediately opened Darwin's eyes to the realities of life at sea, and to a facet of FitzRoy's character that had previously been hidden from him. Being in harbour with no sailing duties on Christmas Day, the British sailors considered it their right to get paralytically drunk, and they were quite prepared to suffer the inevitable consequences. Darwin was appalled; on 25 December he wrote in his diary:

> At present there is not a sober man in the ship:[1] King is obliged to perform duty of sentry, the last sentinel came staggering below declaring he would no longer stand on duty, whereupon he is now

in irons getting sober as fast as he can. – Wherever they may be, they claim Christmas day for themselves, & this they exclusively give up to drunkedness – that sole & never failing pleasure to which a sailor always looks forward.

Monday 26th A beautiful day, & an excellent one for sailing, – the opportunity has been lost owing to the drunkedness & absence of nearly the whole crew. – the ship has been all day in state of anarchy.

So whatever the state of the wind, it would have been impossible to sail on the 26th because of the state of the crew – a virtually hand-picked crew of proven seamen. In swift order, as soon as they were at sea FitzRoy meted out the expected punishment. The Captain's Log for 28 December records the following details:

Disrate William Bruce: Able Seaman to Landsman for breaking his leave, drunkenness and fighting.
Disrate Thos. Henderson: Bosun's mate to Able Seaman for breaking his leave and drunkenness.
Disrate Stephen Chamberlaine: Able Seaman to Landsman for breaking his leave.
Disrate John Wasterham: Captain of Foretop to Able Seaman for breaking leave.
Disrate James Lester: Cooper to Landsman for breaking leave.
John Bruce: 25 lashes for drunkenness, quarrelling and insolence.
David Russel: Carpenter's crew with 34 lashes for breaking his leave and disobedience of orders.
James Phipps: 44 lashes for breaking his leave, drunkenness and insolence.
Elias Davis: 31 lashes for reported neglect of duty.

Other men were put in irons – made to sit for eight hours or more in heavy chains. Darwin, although appreciating 'how absolutely necessary strict discipline is amongst such thoughtless beings as Sailors', was deeply disturbed by all this; he first writes that 'it is an unfortunate beginning' that FitzRoy was obliged 'so early to punish some of our best men', and later records that his thoughts are still 'most unpleasantly occupied with the flogging'. But it would be quite wrong to infer any evidence of a sadistic or brutal streak in FitzRoy's makeup; what the incident shows is that he was prepared to use whatever means were appropriate to carry out his mission, even where the necessity ran counter to his natural instincts. He explains that:

> Individual misconduct, arising out of harbour irregularities, obliged me to have recourse to harsh measures before we had been two days at sea; but every naval officer knows the absolute necessity of what inexperienced persons might think unnecessary coercion, when a ship is recently commissioned. Hating, abhorring corporal punishment, I am nevertheless fully aware that there are too many coarse natures which cannot be restrained without it, (to the degree required on board a ship,) not to have a thorough conviction that it could only be dispensed with, by sacrificing a great deal of discipline and consequent efficiency. 'Certainty of punishment, without severity,' was a maxim of the humane and wise Beccaria; which, with our own adage about a timely 'stitch,' is extremely applicable to the conduct of affairs on board a ship, where so much often depends upon immediate decision, upon instant and implicit obedience.[2]

The key word here is 'efficiency'; above all else, FitzRoy was always concerned with carrying out his duty with maximum efficiency, without always (as we saw in Chapter Two) waiting for official approval from his superiors before changing his plans in the interests of efficiency, even if that meant going against the strict letter of his orders, or dipping into his own, by no means bottomless, pocket to pay for things (like chronometers) that would improve the efficiency of, in this case, his surveying work. The crew obviously shared his views on punishment. It was no more (but no less) than they expected; the scars soon healed, the disrated men gradually earned back their ratings, and the long-term effect amounted to the loss of pay incurred during the disrating (in effect, a fine). FitzRoy's stitch in time served its purpose in establishing that he was not a Captain to be trifled with, and there was little need for floggings on the rest of the five-year voyage. *Beagle* was a happy ship, and many of the crew had volunteered to serve once again under FitzRoy's command; it's just that to modern eyes (as to the twenty-two-year-old gentleman Charles Darwin) the brutality of the disciplinary procedures seems hard to reconcile with the happiness of the ship.

Apart from this key incident, there is no need for us to dwell on the day-to-day routine of sailing the *Beagle*, and it would be tedious to go into repetitive details of the surveying work, which proceeded along the lines of the previous voyage, described in Chapter Two. We shall concentrate on the highlights – particularly those that shed light on FitzRoy's character – but in the constant awareness that this is possible only because FitzRoy's obsession with efficiency ensured the smoothest possible running of that day-to-day routine. Anyone who can make surveying in the stormy seas off South America in a sailing ship the size of the *Beagle* seem dull and routine must indeed have been a superb seaman. A good example of FitzRoy's attention to detail was his standing order that in unknown country nobody was to venture out of sight of the ship unless in a party of at least three – providing one man to stay with the victim in case of an accident, and leaving one free to report back to the ship. This order was disobeyed only once (with fatal results); one consequence is that there is a distinct lack of any exciting tales of lost explorers with which to embroider an account of the voyage. But the *Narrative* needs no such embroidery.

FitzRoy's first major decision had to be made on 6 January 1832, when the *Beagle* dropped anchor at Tenerife, in the Canary Islands. It was intended to make observations to determine the precise longitude of the peak of Mount Teide, which dominates the island, as an aid to navigation. But on being informed that the port authorities had heard reports of an outbreak of cholera in England, and that the ship would be required to spend twelve days at anchor in quarantine before anyone would be allowed ashore, FitzRoy promptly upped anchor and set sail for the Cape Verde islands, a little over 435 miles west of Senegal. This was a particular disappointment to Darwin, who had cherished the thought of visiting the volcanic peak, both for its geological interest and because he suffered badly from seasickness and longed to set foot on solid ground. That pleasure had to wait until the 16th, when the ship anchored in Port Praya, on the island of São Tiago (known to FitzRoy and Darwin as St Jago). During the three weeks that the ship stayed in the Cape Verde Islands while FitzRoy and his team carried out their surveying work (and FitzRoy, as ever, made careful notes on the possibilities for commercial exploitation of the region), Darwin found his feet as a geologist, coming up with the dramatic new idea, based on the evidence of raised beds of material containing marine remains, that the volcanic islands had only recently emerged from the sea, and were still being uplifted by continuing geological activity. He was also (and even more importantly from our present point of view) finding his feet as a diarist, beginning to keep a detailed record not only of his scientific observations but of his thoughts about life in general, and his opinions about his companions on the *Beagle*, particularly FitzRoy. He had never kept a diary before going to sea (although like many of his contemporaries he was an inveterate letter-writer), but learned to do so by following the example of FitzRoy, whose naval training and duties required him to keep an accurate record of everything that happened to the ship.

There was something of a ritual about the process. At noon, FitzRoy took his daily check of the chronometers (the difference between local noon, determined by the height of the sun, and the time in England, recorded on the chronometers, gave the longitude of the ship). This was followed by a meal, at which FitzRoy and Darwin usually ate together, part of Darwin's role as the Captain's gentleman companion. Then, FitzRoy settled down to bring both the ship's log (the formal record of the ship's movements, punishment meted out, and so on) and his own narrative journal up to date. Darwin got into the habit of doing the same, recording both his scientific observations and his own narrative in different books. The record we call his diary he referred to as his journal, or occasionally, when he felt in a more nautical mood, as his Log Book.[3] Even when he was travelling through South America, away from the ship for weeks at a time, he largely maintained the ritual of writing everything down every day in the proper naval manner, following FitzRoy's example.

The historian and Darwin scholar Janet Browne has described the resulting 'plethora of written material' as 'one of the unsung achievements of Darwin's activities on the *Beagle* voyage'.[4] She says that:

> In keeping such copious records, he learned to write easily about nature and about himself. Like FitzRoy, he taught himself to look closely at his surroundings, to make notes and measurements, and to run through a mental checklist of features that ought to be recorded, never relying entirely on memory and always writing reports soon after the event. Like FitzRoy, he became accustomed to thinking about himself as the central character of his text – the captain of his personal natural history travels, the man responsible for planning and executing collecting trips – and accustomed to explaining his course of action, even if it was only a half-hearted justification of money spent on some expedition into the interior. Although this was ordinary practice in naval affairs, it was for Darwin a basic lesson in arranging his thoughts clearly and an excellent preparation for composing logical scientific arguments that stood him in good stead for many years afterwards.

> The daily discipline was severe, and again, he learned from FitzRoy to give it priority. He tagged and numbered specimens, made notes about their location and time of capture, with other details where necessary, and prepared duplicate copies of his lists so that nothing would be irretrievably lost. Darwin had not done this as an amateur specimen-hunter in Edinburgh or Cambridge, and was no doubt advised by Henslow and others on the wisdom of the practice. FitzRoy played an important if unacknowledged role in showing how the undertaking could be carried out.

It is appropriate that through Darwin's diary we can see FitzRoy as others saw him, without having to read between the lines of FitzRoy's own narrative. In addition, we have the insights provided by Darwin's letters home, which flowed in a steady stream, each one written over a period of days or weeks, then sent back to England (with all the other letters from the ship) when the *Beagle* happened to fall in with a homeward-bound vessel. So we find Darwin writing back to his father, shortly after the *Beagle* set sail from the Cape Verde islands on 8 February, that:[5]

> The Captain continues steadily very kind and does everything in his power to assist me. – We see very little of each other when in harbour, our pursuits lead us in such different tracks. – I never in my life met with a man who could endure nearly so great a share of fatigue. – He works incessantly, and when apparently not employed, he is thinking. – If he does not kill himself, he will, during this voyage, do a wonderful quantity of work.

This is the continuing theme in descriptions of FitzRoy – a dedicated professional who worked himself into the ground (in the end, literally to his death) in public service, with little or no thought for his own health, reputation or financial well-being.

From St Jago, the *Beagle* set course for Brazil, crossing the equator on 17 February. The usual horseplay ensued, with Darwin among the victims of King Neptune and his court. The tradition was that everyone 'crossing the line' for the first time would be subjected to various indignities, including being shaved by a sailor acting the part of Neptune and half-drowned in a parody of baptism by total immersion. It was one of the few occasions at sea when discipline was relaxed, an event eagerly anticipated by the men, though not by the Captain. FitzRoy seems to have regarded the ritual, like flogging, as a necessary evil:

> The disagreeable practice alluded to has been permitted in most ships, because sanctioned by time; and though many condemn it as an absurd and dangerous piece of folly, it has also many advocates. Perhaps it is one of those amusements, of which the omission might be regretted. Its effects on the minds of those engaged in preparing for its mummeries, who enjoy it at the time, and talk of it long afterwards, cannot easily be judged of without being an eye-witness.

As FitzRoy recognised, the crossing-the-line activities provided a welcome break from routine at a stage of the voyage when a ship must usually be far from land, with boredom setting in, the possibility of illness, and a long haul ahead to the next port of call. But for the *Beagle*, the next port of call was not far off – on 19 February they arrived at the island of Fernando Noronha, where FitzRoy's instructions from Beaufort required him to make observations, since there was some doubt about the longitude of the island determined by an earlier expedition. Sailing again on the 20th, just eight days later (and two months out of Plymouth) they were able to anchor at Bahia (known today as São Salvador) in the Baia de Todos os Santos (Bay of All Saints) on the mainland of Brazil. They would stay in South American waters for the next three and a half years, and we shall touch on only a few of the highlights of their activities during that time.

By now, Darwin had acquired the nickname 'Philosopher', or 'Philos', being accepted as a member of the ship's company; in turn, he had become used to FitzRoy's moods and the particular danger of annoying him first thing in the morning, when his temper seemed to be on a very short fuse. At that time of day, the Captain made his formal rounds of the ship, inspecting everything, and raging at anything short of perfection. Officers coming on watch had a habit of asking 'whether much hot coffee has been served out this morning?' as a way of enquiring after the state of the Captain's temper; as the other side of the coin, FitzRoy could sometimes abandon himself to moody introspection for hours on end, what Darwin referred to as his 'severe silences'. But none of this prepared the younger man for the explosion that occurred at Bahia.

Darwin, who came from a liberal family which had taken a prominent stand against slavery, was shocked to find that Brazil was a slave state, writing to Henslow that this was a 'scandal to Christian nations'. But FitzRoy seemed unconcerned. Darwin takes up the story:[6]

[FitzRoy] told me that he had just visited a great slave-owner, who had called up many of his slaves and asked them . . . whether they wished to be free, and all answered 'No.' I then asked him, perhaps with a sneer, whether he thought that the answers of slaves in the presence of their master was worth anything. This made him excessively angry, and he said that as I doubted his word, we could not live any longer together. I thought that I should have been compelled to leave the ship; but as soon as the news spread, which it did quickly, as the captain sent for his first lieutenant to assuage his anger by abusing me, I was deeply gratified by receiving an invitation from all the gun-room officers to mess with them. But after a few hours FitzRoy showed his usual magnanimity by sending an officer to me with an apology and a request that I would continue to live with him.

Quick to anger, FitzRoy knew this was a fault, and was almost as quick to make amends; and one of the features of his personality is that he never bore a grudge, even when he was the one who had been wronged.

On 18 March, the *Beagle* sailed south, carrying out a survey of the Abrolhos Islands on her way to Rio. Their course took them past Cape Frio, a high cliff not far from Rio where the *Thetis*, the frigate where FitzRoy had once served as a Lieutenant, had run aground and sunk, with considerable loss of life, in December 1830. FitzRoy describes the tragedy in detail in his *Narrative*, explaining how the combination of bad weather and a previously unknown current had driven the ship off course at night and left the Captain with no route of escape. In offering an explanation of how such a disaster can have occurred in a well-found ship so close to port, he points out the differences between a Royal Navy vessel and a merchantman:

As in the case of the Thetis, an English man-of-war may incur risk in consequence of a praiseworthy zeal to avoid delaying in port, as a merchant-ship would probably be obliged to do, from her being unable to beat out against an adverse wind, and, like that frigate, may be the first to prove the existence of an unsuspected danger.

Those who never run any risk; who sail only when the wind is fair; who heave to when approaching land, though perhaps a day's sail distant; and who even delay the performance of urgent duties until they can be done easily and quite safely; are, doubtless, extremely prudent persons: – but rather unlike those officers whose names will never be forgotten while England has a navy.

And there we have Robert FitzRoy's philosophy of life, and the reason why his name is one of those that have not been forgotten.

We have already noted how, when the *Beagle* arrived in Rio harbour on 4 April 1832, FitzRoy found that 'the charm of novelty being gone', the scene did not live up to his memories of his first arrival there, on board the *Owen Glendower* in 1819. But while the teenage Midshipman had had time to admire the scenery, the Captain had other things to concentrate on as they entered harbour. In the *Narrative*, he merely reports that 'we shortened sail under the stern of our flag-ship'; it is Darwin who tells us that in a typical demonstration of the efficiency of his ship, FitzRoy chose to enter harbour carrying every stitch of canvas that the *Beagle* could carry so that:

We came, in first rate style, alongside the Admirals ship, & we, to their astonishment, took in every inch of canvas & then immediately set it again: A sounding ship[7] doing such a perfect mæneuovre with such certainty & rapidity, is an event hitherto unknown in that class. – It is a great satisfaction to know that we are in such beautiful order & discipline.

On another occasion, when the whole fleet had been exercising under the orders of the flagship, the *Beagle* excelled to such an extent that, as Darwin wrote home to his sisters:

The commanding officer says we need not follow his example, because we do everything better than his great ship. I begin to take a great interest in naval points, more especially now, as I find they all say we are the No. 1 in South America. I suppose the captain is a most excellent officer. It was quite glorious to-day how we beat the *Samarang* in furling sails. It is quite a new thing for a 'sounding ship' to beat a regular man-of-war, and yet the *Beagle* is not at all a particular ship; Erasmus will clearly perceive it when he hears that in the night I have actually sat down in the sacred precincts of the quarterdeck.

The Captain was indeed 'a most excellent officer', and the 'hot coffee' had been poured to good effect. But even the happy, well-disciplined and efficient crew of the *Beagle* can hardly have been delighted when FitzRoy found that there was a difference 'exceeding four miles of longitude' between the distance from Bahia to Rio measured by his chronometers and the distance measured by an earlier French expedition, under Baron Roussin. Although entirely happy that his own observations were correct, he:

> resolved to return to Bahia, and ascertain whether the Beagle's measurement was incorrect. Such a step was not warranted by my instructions; but I trusted to the Hydrographer for appreciating my motives, and explaining them to the Lords of the Admiralty. In a letter to Captain Beaufort, I said, 'I have not the least doubt of our measurement from Bahia; but do not think that any other person would rely on this one measure only, differing widely, as it does, from that of a high authority – the Baron Roussin. By repeating it, if it should be verified, more weight will be given to other measures made by the same instruments and observers.'

Repeat it they did, and it was verified, although on the doubling back three members of the crew died of a fever picked up in Rio. The fever seems to have been malaria, and it is worth reporting FitzRoy's description of the incident in detail, because it highlights the care he tried to take to look after his crew, and how he was always seeking to improve the understanding of natural hazards for the benefit of others. The fact that nobody at that time had any idea that malaria was transmitted by mosquitoes in no way diminishes FitzRoy's efforts in this regard.

> We sailed with the ebb-tide and sea-breeze, cleared the port before the land-wind rose, and when it sprung up steered along the coast towards Cape Frio. Most persons prefer sailing from Rio early in

the morning, with the land-wind; but to any well-manned vessel, there is no difficulty whatever in working out of the port during a fresh sea-breeze, unless the flood-tide should be running in strongly.

On this passage one of our seamen died of a fever, contracted when absent from the Beagle with several of her officers, on an excursion to the interior part of the extensive harbour of Rio de Janeiro. One of the ship's boys, who was in the same party, lay dangerously ill, and young Musters seemed destined to be another victim to this deadly fever.

It was while the interior of the Beagle was being painted, and no duty going on except at the little observatory on Villegagnon Island, that those officers who could be spared made this excursion to various parts of the harbour. Among other places they were in the river Macacu, and passed a night there. No effect was visible at the time; the party returned in apparent health, and in high spirits; but two days had not elapsed when the seaman, named Morgan, complained of headache and fever.

The boy Jones and Mr. Musters were taken ill, soon afterwards, in a similar manner; but no serious consequences were then apprehended, and it was thought that a change of air would restore them to health. Vain idea! they gradually became worse; the boy died the day after our arrival in Bahia; and, on the 19th of May, my poor little friend Charles Musters, who had been entrusted by his father to my care and was a favourite with every one, ended his short career.

My chief object in now mentioning these melancholy facts is to warn the few who are not more experienced than I was at that time, how very dangerous the vicinity of rivers may be in hot climates. Upon making more inquiry respecting those streams which run into the great basin of Rio de Janeiro, I found that the Macacu was notorious among the natives as being often the site of pestilential malaria, fatal even to themselves. How the rest of our party escaped, I know not; for they were eleven or twelve in number, and occupied a day and night in the river. When they left the ship it was not intended that they should go up any river; the object of their excursion being to visit some of the beautiful islets which stud the harbour. None of us were aware, however, that there was so dangerous a place as the fatal Macacu within reach. I questioned every one of the party, especially the second lieutenant and master, as to what the three who perished had done different from the rest; and discovered that it was believed they had bathed during the heat of the day, against positive orders, and unseen by their companions; and that Morgan had slept in the open air, outside the tent, the night they passed on the bank of the Macacu.

As far as I am aware, the risk, in cases such as these, is chiefly encountered by sleeping on shore, exposed to the air on or near the low banks of rivers, in woody or marshy places subject to great solar heat. Those who sleep in boats, or under tents, suffer less than persons sleeping on shore and exposed; but they are not always exempt, as the murderous mortalities on the coast of Africa prove. Whether the cause of disease is a vapour, or gas, formed at night in such situations, or only a check to perspiration when the body is peculiarly affected by the heat of the climate, are questions not easy to answer, if I may judge from the difficulty I have found in obtaining any satisfactory information on the subject. One or two remarks may be made here, perhaps. – the danger appears to be incurred while sleeping; or when over-heated; not while awake and moderately cool; therefore we may infer that a check to the perspiration which takes place at those times is to be guarded against, rather than the breathing of any peculiar gas, or air, rising from the rivers or hanging over the land, which might have as much effect upon a person awake, as upon a sleeper. Also, to prevent being chilled by night damp, and cold, as well as to purify the air, if vapour or gas should indeed be the cause of fever, it is advisable to keep a large fire burning while the sun is below the horizon. But the subject of malaria has been so fully discussed by medical men, that even this short digression is unnecessary.

Here we see both the human side of FitzRoy, caring for his men and upset at the death of little Charles Musters, and FitzRoy the scientist, puzzling over the causes of the disease in a logical fashion. Back in Rio, on the night of 3 June, FitzRoy established that the error in the French measurements stemmed from an incorrect position on their charts for the Abrolhos Islands. The results were communicated to the French through official channels, but FitzRoy never received any response from the Baron. The success of this series of observations, however:

> much increased my own confidence in that simple method of ascertaining differences of longitude, and tended to determine my dependence upon a connected chain of [chronometrical] meridian distances, in preference to any other mode of finding the precise longitude.

Darwin was spared the repetition of this part of the voyage, staying in Rio with (among others) the ship's artist, Augustus Earle, and Fuegia Basket. Although he was busy with his botanical, zoological and geological studies of the surrounding region, this also gave him time to reflect on the voyage so far, and to write at length to his sister Caroline, elaborating not just on his adventures but his thoughts on FitzRoy:

> As far as I can judge, he is a very extraordinary person. I never before came across a man I could fancy being a Napoleon or a Nelson. I should not call him clever, yet I feel convinced nothing is too great or too high for him. His ascendancy over everybody is quite curious: the extent to which every officer and man feels the slightest rebuke or praise, would have been before seeing him, incomprehensible. It is very amusing to see all hands hauling at a rope, they not supposing him on deck, and then observe the effect when he utters a syllable; it is like a string of dray horses, when the waggoner gives one of his awful smacks. His candour and sincerity are to me unparalleled, and using his own words his 'vanity and petulance' are nearly so. I have felt the effects of the latter: but the bringing into play the former ones so forcibly makes one hardly regret them. His greatest fault as a companion is his austere silence: produced from excessive thinking: his many good qualities are

great and numerous: altogether he is the strongest marked character I ever fell in with.

With everyone back on board,[8] the *Beagle* finally left Rio heading south on 5 July 1832. She was sped on her way by three hearty cheers from HMS *Warspite*. 'Strict etiquette,' writes FitzRoy, 'might have been offended' at such a salute to such a tiny ship, or to any vessel unless she was going out to meet an enemy or returning from a victory: 'But although not about to encounter a foe, our lonely vessel was going to undertake a task laborious, and often dangerous, to the zealous execution of which the encouragement of our brother-seamen was no trifling inducement.'

FitzRoy soon had another chance to demonstrate the characteristics Darwin remarks upon. After calling at Montevideo, at the mouth of the River Plate, he decided to visit Buenos Aires, across the wide Plate estuary, to inform the government there of his activities in the area and to collect any relevant charts. In his *Narrative*, FitzRoy passes over what happened next (on 2 August) in a couple of sentences; Darwin's diary provides more insight:

> On entering the outer roadstead, we passed a Buenos Ayres guardship. When abreast of her she fired an empty gun; we not understanding this sailed on, and in a few minutes another discharge was accompanied by the whistling of a shot over our rigging. Before she could get another gun ready, we had passed her range.

But at the harbour proper, nobody was allowed ashore, and they were informed that they must undergo quarantine for cholera. Nothing they could say would induce the port authorities to relent, even though it had been seven months since the ship left England, so FitzRoy upped anchor and headed back to Montevideo.

> We then loaded and pointed all the guns on one broadside,[9] and ran down close along the guard-ship. Hailed her, and said, that when we again entered the port, we would be prepared as at present

and if she dared to fire a shot we would send our whole broadside into her rotten hulk.

FitzRoy remembered it slightly differently. In an official report to Beaufort, dated 16 August 1832, he says that his words were: 'If you dare to fire another shot at a British man-of-war you may expect to have your hulk sunk, and if you fire at *this* vessel, I will return a broadside for every shot!' And no doubt he would have.

The excitement wasn't over. *Beagle* returned to Montevideo, where FitzRoy reported to the commander of the frigate *Druid*, who sailed immediately to make an official protest in Buenos Aires, where he duly received a full apology for the way *Beagle* had been treated. As soon as the *Druid* disappeared below the horizon, the chief of police came on board, requesting FitzRoy's help in maintaining order in the town, which was being threatened by what FitzRoy refers to as 'turbulent mutineers'. The presence of fifty well-armed men from the *Beagle*, led (of course) by FitzRoy himself and garrisoning the town fort, was enough to restore order in the twenty-four hours it took for more troops to arrive from the interior. No shots were fired, but the incident serves as a reminder of how turbulent conditions were in even the most civilised parts of South America at the time. 'There certainly is a great deal of pleasure in the excitement of this sort of work,' wrote Darwin; 'quite sufficient to explain the reckless gaiety with which sailors undertake even the most hazardous attacks. Yet as time flies, it is an evil to waste so much in empty parade.'

It was about this time that Darwin began to make use of his time ashore to collect fossils, much to the initial amusement of FitzRoy and the crew of the *Beagle*, although as FitzRoy acknowledges, 'the cargoes of apparent rubbish which he frequently brought on board . . . have since proved to be most interesting and valuable remains of extinct animals'. And all of which, thanks to FitzRoy's friendship, was shipped back to Britain in Navy vessels at no expense to Darwin, in packing cases made by the *Beagle*'s carpenter. For the next few months, the *Beagle* was engaged in surveying work, with her main base in Montevideo.

FitzRoy, like Darwin, was aware of the passage of time. His instructions required him not only to survey the coastline of southern

South America itself (as if that were not enough) but to include a diversion to the Falkland Islands, out in the Atlantic between 51° and 53° south and 57° to 62° west, a little less than 310 miles to the east of Patagonia. There, Britain was just establishing her authority, for the political and trade considerations that were the reason for the *Beagle*'s presence in South America, and the islands were increasingly being used as a stepping stone on the route to and around Cape Horn. It was impossible for a single ship to carry out all the tasks laid down by Beaufort in his instructions, and even the *Beagle* was too large to safely explore some of the dangerous shoals along the coastline south of the Plate, so FitzRoy did the logical thing – he hired two tiny vessels, graced with the name schooners from their rigging, but displacing only fifteen and nine tons respectively, from a Mr Harris. The idea, which proved entirely successful, was that in the coming southern summer the *Paz* and *Liebre*, manned by members of the *Beagle*'s crew, would continue the surveying along the eastern coast of South America, while *Beagle* headed for the Falklands. But there was a snag:

> One serious difficulty, that of my not being authorized to hire or purchase assistance on account of[10] the Government, I did not then dwell upon, for I was anxious and eager, and, it has proved, too sanguine. I made an agreement with Mr. Harris, on my own individual responsibility, for such payment as seemed to be fair compensation for his stipulated services, and I did hope that if the results of these arrangements should turn out well, I should stand excused for having presumed to act so freely, and should be reimbursed for the sum laid out, which I could so ill spare. However, I foresaw and was willing to run the risk, and now console myself for this, and other subsequent mortifications, by the reflection that the service entrusted to me did not suffer.

Here, again, is an expression of the creed FitzRoy lived by. He would not shrink from any course of action, even though he foresaw the risk to himself, provided that 'the service entrusted to me did not suffer'. And it was duty, albeit a self-imposed duty, that now took the *Beagle* south to Tierra del Fuego to return the Fuegians to their homeland before commencing the survey work around the Falklands.

It may seem strange that we have hardly mentioned the Fuegians in our account of the voyage so far, but the reason is that up to this point they are hardly mentioned by FitzRoy, or Darwin, or in any of the other surviving documents written by those on board. They seem to have been taken for granted, part of the furniture, even though their presence was one of the principal reasons for the voyage. With so many of her people away with the 'schooners', the *Beagle* was undermanned, and for this leg of the voyage FitzRoy borrowed an extra watchkeeping officer, an old friend, Robert Hammond, from the *Druid*, where he was serving as Mate (that is, a passed Lieutenant waiting for a vacancy). They sailed from Montevideo on 27 November 1832, with eight months' supplies, stopping off, en route to Cape Horn, for a rendezvous with Lieutenant Wickham, in command of the schooners.

On 17 December, having been frustrated in earlier landing attempts by bad weather, they dropped anchor in a sheltered harbour known as Good Success Bay, on the southeastern tip of Tierra del Fuego. It was here that Darwin and the other first-time members of the ship's complement had their first encounter with 'wild' Fuegians, and were struck by the contrast with their civilised shipboard counterparts, especially the now elegant Jemmy Button, who, says Darwin, 'used always to wear gloves, his hair was neatly cut, and he was distressed if his well-polished shoes were dirtied. He was fond of admiring himself in a looking-glass.' Both York Minster and Jemmy Button laughed and poked fun at the wild savages they were now confronted with, and claimed not to understand their language, refusing to act as interpreters. In Jemmy's case, it seems he really had forgotten his own language; but York promptly gave himself away when an old man made a comment about York's beard (the wild Fuegians abhorred facial hair, and plucked it out with mussel shells used as tweezers) which he immediately repeated to the Captain. The missionary Richard Matthews was sanguine about his first encounter with the people he was supposed to live among. FitzRoy says that he 'did not appear to be at all discouraged by a close inspection of these natives. He remarked to me, that "they were no worse than he had supposed them to be".'

But first, FitzRoy had to get the Fuegians, and Matthews, to the proposed site for his settlement. The plan was to round Cape Horn and leave York and Fuegia Basket (accepted by everyone as his future 'wife') with their own people, then to head back east through the Beagle Channel, surveying as they went (FitzRoy was *always* surveying as he went), dropping off Matthews and Jemmy to found the mission station before proceeding to the Falklands. They left Good Success Bay on the 21st, and rounded Cape Horn comfortably on the 22nd, but were then forced by bad weather to take shelter, spending Christmas Day at anchor – just as they had been forced by bad weather to spend the previous Christmas at anchor, in Plymouth, with such uncomfortable consequences for some of the crew. By the 31st, 'tired and impatient at the delay', FitzRoy put to sea again as soon as the wind from the west eased sufficiently that the *Beagle* could make progress into it, but the gales continued and their progress was painfully slow. On 11 January 1833, they sighted the promontory that York Minster was named after, but FitzRoy's hopes of finding an anchorage were dashed by the increasing strength of the wind. As the strength of the storm increased (this was midsummer, remember) the *Beagle* barely held her own, until shortly after 1 p.m. on the 13th:

I was anxiously watching the successive waves, when three huge rollers approached, whose size and steepness at once told me that our sea-boat, good as she was, would be surely tried. Having steerage way, the vessel met and rose over the first unharmed, but, of course, her way was checked; the second deadened her way completely, throwing her off the wind; and the third great sea, taking her right a-beam, turned her so far over, that all the lee bulwark, from the cat-head to the stern davit, was two or three feet under water.

For a moment, our position was critical; but, like a cask, she rolled back again, though with some feet of water over the whole deck. Had another sea then struck her, the little ship might have been numbered among the many of her class which have disappeared: but the crisis was past – she shook the sea off through her ports, and was none the worse – excepting the loss of a lee-quarter boat, which, although carried three feet higher than in the former voyage (1826–1830), was dipped under water, and torn away.

But how did the water escape through her ports? According to Sulivan, FitzRoy had a standing order that the gun ports should be secured, believing that a well-handled ship would be more at risk from water coming in through the ports than in need of an outlet for water that had got on to the deck. Sulivan worried about this during the storm and told the carpenter to stand by with a handspike in case of need. According to his account, the combined efforts of himself and the carpenter were crucial in hammering open one of the ports and allowing water to escape before the further wave that would deliver the coup de grâce could hit the ship. As Darwin refers to the decks being 'filled with sea' before 'at last the ports were knocked open & she again rose buoyant to the sea', there may be something in this story – including a hint, from his failure to mention Sulivan's role in his *Narrative*, that FitzRoy was reluctant to acknowledge that he might have made a mistake.

Whatever, the storm, one of the worst ever encountered near Cape Horn in summer, eased sufficiently for the *Beagle* to make a safe anchorage – at which point York Minster, either chastened by his experience of the storm or with some deeper motive, announced that he would be happy if he and Fuegia were left to settle with Jemmy and Matthews, rather than being returned to their own tribe. FitzRoy was delighted at the prospect of the three civilised Fuegians settling together (quite apart from this removing the need to waste more time trying to reach York Minster's own territory), and 'little thought how deep a scheme master York had in contemplation'.

With the *Beagle* in a safe anchorage at the eastern end of the Beagle Channel, plans were made for a boat party to proceed through the narrows to establish the settlement. The ship's yawl was decked over so that it could carry more cargo, the sight of which prompted Darwin to write in his diary:

The choice of articles showed the most culpable folly and negligence. Wine glasses, butter-bolts, tea-trays, soup tureens, mahogany dressing-case, fine white linen, beaver hats and an endless variety of similar things, shows how little was thought about the country where they were going to. The means absolutely wasted on such things would have purchased an immense stock of really useful articles.

The yawl, laden with this cargo (together with, it has to be said, useful things like gardening tools and seeds), would be towed by the three remaining whaleboats when the wind was against them, on an expedition that would last for twenty-three days. As well as Matthews (who 'showed no sign of hesitation or reluctance') and the Fuegians, the party under FitzRoy was made up of 'Darwin, Bynoe, Hamond, Stewart and Johnson, with twenty-four seamen and marines'. Departing on 19 January 1833, travelling by day and laying up under the stars by night just as on the first *Beagle* expedition, they reached Jemmy's people without undue alarms, arriving at his ' "own land", which he called Woollya' on the 23rd. Jemmy's 'own people' were the tribe known to FitzRoy as Tekeenica; in a delicious and classic misunderstanding, in their language 'teke uneka' actually means 'I do not understand you', the response they had made when he asked them what tribe they were!

Although it turned out that Jemmy had indeed forgotten most of his own language, what seemed to be friendly relations with the natives were soon established, with many of Jemmy's relations (including two brothers dubbed Tommy Button and Harry Button by the sailors) appearing. Assured by York that Jemmy's brother was 'very much friend' and that the location was 'very good land', FitzRoy picked out a pleasant sheltered spot near a brook of excellent water, and with a boundary line marked out the party set about constructing three large wooden wigwams, one each for Matthews and Jemmy, the other for York and Fuegia, and planting a vegetable garden. Word about all this activity soon spread, and by the 26th Fuegians from the tribe known to FitzRoy as Yapoos had arrived in force, so there were about 300 natives close by the encampment of just thirty men from the *Beagle*. FitzRoy took the opportunity of arranging firing practice that evening, 'with the three-fold object of keeping our arms in order – exercising the men – and aweing, without frightening, the natives'. The demonstration was more effective than he had planned; next day, all the natives, including Jemmy's family, packed up their belongings, got into their canoes, and left. FitzRoy decided 'to take the opportunity of their departure to give Matthews his first trial of passing a night at the new wigwams'. Under cover of the gathering dusk, that evening the larger and more valuable items of his equipment were transferred to the settlement and hidden in his makeshift cabin, or buried in a box under the floor. 'Matthews was steady, and as willing as ever; neither York nor Jemmy had the slightest doubt of their being all well-treated,' so FitzRoy determined to leave Matthews to fend for himself for a few days. In fact, to start with he left him for only one night, sailing back to Woollya the next morning, his own anxiety 'increased by hearing the remarks made from time to time by the rest of the party, some of whom thought we should not see him alive', so that it was 'with no slight joy that I caught sight of him, as my boat rounded a point of land, carrying a kettle to the fire near his wigwam'.

Nothing had happened to dampen Matthews's enthusiasm, and although some natives had returned at daybreak, these included one of Jemmy's brothers, they were completely friendly, and Jemmy himself assured the Captain that 'his own family would come in the course of the day, and that the "bad men", the strangers, were all gone away to their own country'. Short of abandoning the whole project on the spot, there was nothing for it but to leave Matthews for a longer trial. The yawl and one whaleboat were sent back to the *Beagle*, while FitzRoy took the other two whaleboats (and Darwin) off surveying to the westward. It was on this expedition that a large expanse of water was named Darwin Sound, 'after my messmate, who so willingly encountered the discomfort and risk of a long cruise in a small loaded boat'. On the way back to Woollya, however, they were perturbed to come across a group of natives who were heavily painted and decorated with, among other things, scraps of red cloth and pieces of ribbon, while one woman was wearing a linen garment that had belonged to Fuegia Basket. In a state of high anxiety about the fate of Matthews and his party, they pressed on, arriving at the settlement on the morning of 6 February. On the beach were several canoes, and 'as many natives seemed to be assembled as were there two days before we left the place.

All were much painted, and ornamented with rags of English clothing, which we concluded to be the last remnants of our friends' stock.' But to FitzRoy's great relief, 'Matthews appeared, dressed and looking as usual. After him came Jemmy and York, also dressed and looking well: Fuegia, they said, was in a wigwam.'

FitzRoy took Matthews on board his boat, and moved a little offshore to hear his story free from interruption. It made grim listening. Everything had seemed to be going well until three days after FitzRoy's departure, when several canoes full of strangers had arrived; from that time on Matthews had been subject to continual harassment, stopping just short of actual violence (except for one occasion when his head had been forcibly held down to the ground, as if in a contemptuous demonstration of his physical weakness and powerless position). At the time of FitzRoy's return the Fuegians had been forcibly plucking out the hairs of his beard, using their usual mussel shell tweezers – but although this is often interpreted as a sign of hostility towards the stranger, it could just as well be argued that they were trying to assimilate him into their community by making him look more like them. Either way, it was not a pleasant experience. Both his own and Jemmy's possessions were plundered, though York and Fuegia suffered less; the vegetable garden was trampled underfoot in spite of Jemmy's attempts to explain what it was for.

It was clear that Matthews could not remain among such savages. The immediate problem was how to get his remaining possessions and the hidden valuables away with him; FitzRoy achieved this by telling his men to spread out as if they intended a lengthy stay, and when the natives were lulled into this belief he quickly moved to collect the valuables and transfer them to the boats. Anxious to do his best for Jemmy, York and Fuegia, he distributed presents of axes, saws, gimlets, knives and nails, 'then bade Jemmy and York farewell, promising to see them again in a few days, and departed from the wondering throng assembled on the beach'.

The party returned to the *Beagle*, where all was well, surveying had been completed and the ship refitted ready for the next leg of her work. After several days investigating bays and passages in the eastern part of Tierra del Fuego, on 14 February FitzRoy returned to Woollya to keep his promise to Jemmy. There were only a few natives there, and Jemmy, York and Fuegia were all in their civilised clothing, everything seemed in order and vegetables were sprouting in the remains of the garden. York was busy building a large canoe, and Jemmy said 'that he should get on very well now that the "strange men" were driven away'. FitzRoy 'left the place, with rather sanguine hopes of their effecting among their countrymen some change for the better'. Only a blind optimist could ever have hoped that Matthews could have single-handedly acted as a civilising influence upon such savages; but FitzRoy, though he had clearly had definite doubts about this part of the project all along, still remained optimistic that the presence of York, Jemmy and Fuegia in their midst might act to uplift the natives.

> I hoped that through their means our motives in taking them to England would become understood and appreciated among their associates, and that a future visit might find them so favourably disposed towards us, that Matthews might then undertake, with a far better prospect of success, that enterprise which circumstances had obliged him to defer, though not to abandon altogether.

That optimism was only destroyed in stages over many months.

Furious winds delayed the departure of the *Beagle* from Tierra del Fuego until 26 February, so it was only at the beginning of March 1833 that they arrived in the Falklands for the first time. The political status of the islands was vague at the time. They had been claimed at various times by Spain, France, Britain and Argentina. The latest attempt at settlement had been made by a group from South America led by one Lewis Vernet, a German by birth, whose activities were formally sanctioned by the government in Buenos Aires in 1828, in the face of protests from the British, who said that the Buenos Aires government had no right to the islands. Following a dispute with North American sealers, Vernet's settlement was largely destroyed by the crew of the United States corvette *Lexington*, whose Captain, Silas Duncan, took it upon himself to carry several of the leading settlers off to the mainland of South America in February 1832. While the South and North Americans were debating the issue, the British sent HMS *Clio* and

HMS *Tyne* to assert their right to sovereignty. They arrived in the Falklands on 2 January 1833, leading the small garrison from Buenos Aires to slip away quietly in their armed schooner. Having literally shown the colours, the two ships sailed off again, leaving no authority except for a Union Flag in the care of an Irishman who had been Vernet's storekeeper, with instructions to hoist it every Sunday and to make sure it was shown if any ships came to the islands. So it was with understandable caution that FitzRoy, who had been out of communication for months, approached the islands less than two months later, and decided to anchor well out from shore when he saw a French flag flying near some tents at a bay known as Johnson Cove.

The French, however, turned out to be shipwrecked mariners, from whom FitzRoy learned about the current political situation. He agreed to take some of them to Montevideo, and to help find passage for the rest, but was shocked to find that although the weather was now fine and their ship was lying on a sandy shore, even though they had been able to make themselves comfortable in large tents made from her sails no attempt had been made at salvage, even though 'all the stores and provisions, if not the ship herself, might have been saved by energetic application of proper means soon after she was stranded. When I saw her it was not too late, but I had too many urgent duties to fulfil to admit of my helping those who would not help themselves.' Again, the true FitzRoy speaks!

Next day, 2 March, as surveying operations began, FitzRoy learned of the arrival of a Mr Brisbane, Vernet's agent and partner, who had come to pick up the pieces of the settlement wrecked by the *Lexington*. Brisbane acted in a private capacity, FitzRoy stresses, with no hint here of a renewal of the Argentine claim to the islands. He came in a merchant schooner, the *Rapid*, which, belying her name, had taken fourteen days to reach the Falklands from Buenos Aires. With the heavy irony of an old Cape Horn hand, FitzRoy describes how 'no sooner had Mr. Brisbane landed than the master and crew of the Rapid hastened to make themselves drunk, as an indemnification for the fatigues of their exceedingly long and hazardous voyage'. It was also on this day that the value of FitzRoy's orders about never leaving sight of the ship except in parties of three or more was grimly brought home.

The ship's Clerk, Edward Hellyer, had walked about a mile from the ship, initially in the company of a Frenchman who then left him, to shoot duck. When he did not return a search was organised and his body was found under the water, tangled in kelp. His discharged shotgun lay on the shore with his clothes and watch, and a dead duck lay just out of his reach. It was clear that he had shot the duck, gone into the water to retrieve it, got tangled in the kelp and drowned. The death hit FitzRoy hard:

> To me this was as severe a blow as to his own messmates; for Mr. Hellyer had been much with me, both as my clerk and because I liked his company, being a gentlemanly, sensible young man. I also felt that the motive which urged him to strip and swim after the bird he had shot, was probably a desire to get it for my collection.

But there was no time to brood over such matters. In the middle of March, with autumn setting in, a sealing schooner, the *Unicorn*, arrived at the islands. The Master and part owner, William Low, had experienced such foul weather that his six months' sealing expedition had been a disaster, and he was financially ruined. The wheels turned quickly in FitzRoy's head.

> At this time I had become more fully convinced than ever that the Beagle could not execute her allotted task before she, and those in her, would be so much in need of repair and rest, that the most interesting part of her voyage – the carrying a chain of meridian distances around the globe – must eventually be sacrificed to the tedious, although not less useful, details of coastal surveying . . .
>
> I had often longed for a consort, adapted for carrying cargoes, rigged so as to be easily worked with few hands,[11] and able to keep company with the Beagle; but when I saw the Unicorn, and heard how well she had behaved as a sea-boat, my wish to purchase her was unconquerable.

She was, indeed, a fine vessel of 170 tons, copper-bottomed and, in the expert opinion of the carpenter, Jonathan May, in good condition. Low had the authority to act on his partners' behalf, and FitzRoy bought the *Unicorn* for the equivalent of £1,300 (about £90,000 in today's money). The schooner had an interesting past. She had been built in Rochester, on the Medway, as a private yacht, for about £6,000, but shortly afterwards she was armed and used by Lord Thomas Cochrane (the archetype for several fictional sea heroes, including Hornblower and Aubrey) in the Mediterranean. On service in South American waters, she had been taken by the Brazilians, sold to the British Consul in Montevideo (who used her as a private yacht on a trip to England and back), then fitted out for the disastrous sealing expedition. FitzRoy was able to obtain all the materials he needed to fit her out from salvage from shipwrecks in the Falklands, paying about a third of the price he would have expected in Montevideo. Finally, 'to keep up old associations' he renamed the schooner *Adventure*.

Aware how far he had exceeded his official instructions, even if for the good of the service, FitzRoy wrote to Beaufort on 10 May 1833 begging for his active support:

> I am most anxious to hear who is to pay for the two little craft 'Paz' and 'Liebre', – whether His Majesty or His Majesty's humble servant . . . Now *pray fight my battle* & get me *twenty super-numerary Seamen* for the Beagle – *fifteen* AB & *five first class* ratings – it will save my pocket *so very much*, & have them you know I must, either for his *Majesty* or *myself.* I feel as if we could now get on fast again, & much more securely, by having so fine a craft [as *Adventure*] to carry our *luggage, provisions, boats*, &c &c. I mean to make her a regular '*Lighter*'.[12]

But their Lordships were unimpressed – a minute in the Admiralty Record Office referring to the *Paz* and *Liebre* carries the comment: 'Do not approve of hiring vessels for this service, and therefore desire that they may be discharged as soon as possible.' And at this time the political complexion of the Admiralty was not one where FitzRoy's 'interest' carried any weight – the long-standing Tory administration had been replaced by a Whig government, and FitzRoy came from a prominent Tory family. The eternal optimist was storing up even more trouble for himself than he might have guessed.

But all of that lay far in the future when *Adventure* sailed from the Falklands on 4 April 1833, followed by the *Beagle* two days later. Having rendezvoused with the rest of his party, FitzRoy set about reorganising things with the *Paz* and the *Liebre* returned to their owner (FitzRoy had to meet the cost of their hire, £1,680, out of his own pocket), and decided to have the bottom of the *Adventure*, now commanded by Wickham, entirely re-coppered (again, at his own expense) in anticipation of her voyage in company with the *Beagle* through the Pacific Ocean. For the rest of the year, surveying work continued on the east coast of South America, with Darwin taking the opportunity to make four lengthy trips into the interior which were to prove of great value in developing his ideas on geology and natural history.

Among several changes in the personnel of the *Beagle* during these months, the official artist Augustus Earle had left the ship because of ill health, and been replaced by Conrad Martens, while Benjamin Bynoe, who was proving a very satisfactory surgeon, acquired an assistant, William Kent. Any guilt we may have felt in skipping over FitzRoy's activities during these months is eased by his own reference to the 'tedious, although not less useful, details of coastal surveying'; it was business as usual, and we can pick up the story again on 6 December 1833, when 'with a supply of provisions and coals, sufficient for at least nine months, the Beagle and her tender sailed together from Monte Video', heading yet again for Tierra del Fuego and enjoying their second Christmas away from home along the way. There is, though, one document from this period that shows FitzRoy's interest, even at this early stage, in weather forecasting. A set of sailing directions for the Rio de la Plata, dated 30 November 1833, has an addition made by FitzRoy at a later date: 'South East gales are foretold by a high barometer, with cloudy threatening weather, and lightning: – a red sky at sunrise, and a high river with a strong current running up.'

After Christmas, celebrated at anchor in the bay known as Port Desire, with games ashore in the afternoon, *Beagle* left *Adventure* to complete some alterations to her mast and rigging, and carried on the slow task of surveying her way southwards towards Cape Horn. The most significant incident occurred early in the voyage, on 4 January 1834, when she struck a submerged rock 'so as to shake her fore and aft'. But there were no leaks, and after an inspection of the ship's bottom FitzRoy decided that any damage could be repaired at a later date.[13] *Adventure*, meanwhile, was heading off to the Falklands on her own surveying duties.

It wasn't until 5 March, just over a year since they had last seen Jemmy Button and his companions, that the *Beagle* dropped anchor at Woollya. The 'settlement' was a forlorn sight. The wigwams were empty, and seemed to have been so for months, though undamaged, and the vegetable garden had been trampled but still contained some turnips and potatoes which were pulled up and later eaten at the Captain's table – proof that they could be grown in the region. But there were no natives, and:

> An anxious hour or two passed, after the ship was moored, before three canoes were seen in the offing, paddling hastily towards us, from the place now called Button Island. Looking through a glass I saw that two of the natives in them were washing their faces, while the rest were paddling with might and main: I was then sure that some of our acquaintances were there, and in a few minutes recognized Tommy Button, Jemmy's brother. In the other canoe was a face which I knew yet could not name. 'It must be some one I have seen before,' said I, – when his sharp eye detected me, and a sudden movement of his hand to his head (as a sailor touches his hat) at once told me it was indeed Jemmy Button – but how altered!

Jemmy was naked, like his companions, and 'wretchedly thin'; but within half an hour he was clothed, 'sitting with me at dinner in my cabin, using his knife and fork properly, and in every way behaving as correctly as if he had never left us'. His English was as good as ever – to the astonishment of FitzRoy and his officers, rather than him recovering his fluency in his native language, his companions, who included his new wife, his brothers and their wives, had picked up many English words which they used in speaking to him (they even called him 'Jemmy

Button'). The story he had to tell was a simple one of betrayal – several months earlier, York, helped by some of his brothers, had stolen all of Jemmy's possessions while he slept, loaded them into the large (suspiciously large, with hindsight) canoe that FitzRoy had seen him constructing, and paddled off with Fuegia back to his own tribe. FitzRoy was

> quite sure that from the moment of his changing his mind, and desiring to be placed at Woollya, with Matthews and Jemmy, he meditated taking a good opportunity of possessing himself of every thing; and that he thought, if he were left in his own country without Matthews, he would not have many things given to him, neither would he know where he might afterwards look for and plunder poor Jemmy.

But 'poor Jemmy' seemed quite content with the way things had turned out. He was living with his family (including the young wife) at a safer spot than Woollya, and in any case the abandoned wigwams had proved too cold in winter because the Englishmen had built them too tall. Although to the Europeans he was so thin that he looked ill, he replied to FitzRoy's anxious enquiries that he was 'hearty, sir, never better', and that he had no wish to change his way of life. He had 'plenty fruits', 'plenty birdies', 'ten guanaco in snow time', and 'too much fish'. He had two fine otter skins which he had dressed and kept specially as presents for the ship's Bosun, Bennett, and for FitzRoy himself; he had also made two spear heads as presents for Darwin, and gave FitzRoy a bow and a quiver full of arrows to take back for the schoolmaster with whom he had lived in Walthamstow.

As FitzRoy saw it, 'the first step towards civilization – that of obtaining their confidence – was undoubtedly made: but an individual, with limited means, could not then go farther'. The attempt had been made on too small a scale, and there was clearly no point in leaving Matthews to try his luck once again. But:

> I cannot help still hoping that some benefit, however slight, may result from the intercourse of these people, Jemmy, York, and

Fuegia, with other natives of Tierra del Fuego. Perhaps a ship-wrecked seaman may hereafter receive help and kind treatment from Jemmy Button's children; prompted, as they can hardly fail to be, by the traditions they will have heard of men of other lands; and by an idea, however faint, of their duty to God as well as their neighbour.

FitzRoy's hopes were to be cruelly dashed,[14] and at a time when he was already at a low ebb. But that lay far in the future; after another day at Woollya, the *Beagle* took its leave with these optimistic thoughts still uppermost in his mind. The natives were loaded with presents, and Darwin summed up the general mood in his diary:

> Every soul on board was heartily sorry to shake hands with him for the last time. I not now doubt that he will be as happy as, perhaps happier than, if he had never left his own country. Every one must sincerely hope that Captain FitzRoy's noble hope may be fulfilled . . . When Jemmy reached the shore, he lighted a signal fire, and the smoke curled up, bidding us a last and long farewell, as the ship stood on her course into the open sea.

That course took them once again to the Falkland Islands where they arrived on 10 March to find that a great deal had happened in their absence.[15]

When the small Argentine garrison had left the islands following the arrival of the *Clio*, they had left behind some escaped prisoners, mutineers who had rebelled against the harsh conditions they were forced to live under. When *Clio* in her turn departed, leaving no other authority than a flag, in August 1833 these men and a few of the gauchos who handled the wild cattle in the interior of the islands plundered the tiny settlement at Port Louis. They killed five men, including Brisbane, and drove all the cattle and horses off into the interior. The rest of the settlers (thirteen men, three women and two children) escaped to a small islet where they survived on birds' eggs and fish until the arrival of the sealer *Hopeful*, which provided them with food and other necessities but, says FitzRoy, 'could not delay to protect them from the assaults which they anticipated', although he does not tell us why. It was only a month after that that HMS *Challenger* arrived, carrying a Lieutenant Smith and four seamen who had volunteered to establish a formal British presence in the Falklands. The nine rebels headed into the interior. After an abortive search for them, Smith and his men were left, with the addition of six marines to the party, while the *Challenger* proceeded about her duties. By the time FitzRoy arrived, order had been restored, and the ringleader of the murderers had been captured. Smith also had two other men in custody, one said to be an accomplice to the murderers although not active in the plot, and one who had offered to turn King's evidence. FitzRoy agreed to take these three men back to the mainland.

While all this had been going on William Low, the former master of the *Adventure* and a friend of Brisbane, had been on a sealing expedition with four other men. On their return, he felt that his friendship with Brisbane put his life particularly at risk, and they stood off from the islands, eventually meeting up with the *Adventure*, going about her surveying work to the west and unaware of these developments, on 6 February 1834. Now doubly ruined and with no reason to stay in the Falklands, Low offered his services as a pilot.

> They were accepted, provisionally by Lieut. Wickham, and afterwards by me, trusting that the Admiralty would approve of my so engaging a person who, in pilotage and general information about the Falklands, Tierra del Fuego, Patagonia, and the Galapagos Islands, could afford us more information than any other individual, without exception.

Like Low, FitzRoy had no incentive to stay in this grim place and, leaving *Adventure* to complete the required surveying work, the *Beagle* finally left the Falklands early in April, arriving at the mouth of the Rio Santa Cruz in Patagonia on the 13th. FitzRoy wanted to examine the state of the ship's bottom before leaving the Atlantic behind, 'to ascertain how much injury had been caused by the rock at Port Desire, and to examine the copper previous to her employment in the Pacific Ocean, where worms soon eat their way through unprotected planks'.

This was done, in routine fashion, by beaching the ship at high tide, waiting for the tide to go out, and working on the ship's bottom before it came in again. They found that a piece of the false keel under the ship had been knocked off, and a few sheets of copper needed replacing, but May and his team completed everything in one tide, while Martens took the opportunity to make one of his most famous sketches of the *Beagle*.

Now, FitzRoy could carry out a long planned expedition, which would give him his biggest break from the tedious details of coastal surveying during the entire voyage. His predecessor, Captain Stokes, had made a partial exploration of the Rio Santa Cruz by boat, and FitzRoy was determined to go farther up the river, exploring, and in hope of at least catching sight of the Andes. Leaving Sulivan in charge of the *Beagle*, a party of twenty-five, including FitzRoy himself and Darwin, set out in three whaleboats on an expedition that lasted until 8 May. The river was about 1,000 feet wide, and the boats were light enough to be manhauled upstream, against the current of four to six knots; as Darwin emphasises, everyone took their turn at the task, ninety minutes on and ninety minutes off. On a good day they made twenty miles up river, which worked out at about half that distance in a straight line. On the return journey, they covered more than 80 miles a day sailing in the rapidly flowing water.

From our perspective, the trip is interesting not so much because of the element of adventure and exploration, but because it gives FitzRoy the opportunity, in his *Narrative*, to remind us that he was no mean geologist in his own right. It isn't always appreciated, for example, that the copy of the first volume of Charles Lyell's *Principles of Geology* that Darwin carried with him on the voyage was given to him by FitzRoy. As they go into the interior, he comments:

> Is it not remarkable that water-worn shingle stones, and diluvial accumulations, compose the greater portion of these plains? On how vast a scale, and of what duration must have been the action of those waters which smoothed the shingle stones now buried in the deserts of Patagonia.

There is no hint here of a man who believes in the literal truth of a Biblical Flood lasting for less than a year! Later he comments on the apparent antiquity of the river itself:

> Though the bed of the river is there so much below the level of the stratum of lava, it still bears the appearance of having worn away its channel, by the continual action of running water. The surface of the lava may be considered as the natural level of the country, since, when upon it, a plain, which seems to the eye horizontal, extends in every direction. How wonderful must that immense volcanic action have been which spread liquid lava over the surface of a vast tract of country.

In other words, he sees clear evidence of an ancient period of volcanic upheaval, followed by a length of time long enough for the river to have carved a gorge through the layer of lava to its present position – evidence that clearly conflicts with the Biblical interpretation of the age of the Earth, popular in FitzRoy's day, which set the date of the Creation as 4004 BC. It is still not precisely clear just how and why FitzRoy became transformed into a Biblical literalist over the next few years, but these observations alone are enough to show that he was not specifically looking for evidence in support of the Biblical account of Creation during the voyage, and that he can hardly have been in dispute with Darwin over the interpretation of their observations at that time. Incidentally, FitzRoy gave a talk about these geological studies to the Royal Geographical Society on his return home.

But all this was an aside from his proper work. On 12 May 1834 the *Beagle* was back at sea, and on the 23rd they met up with the *Adventure* off the eastern shores of Tierra del Fuego. By 3 June, with all the local surveying work completed, both vessels were moored in Port Famine, ready for the passage through the maze of channels to the Pacific. Surely significantly, no attempt was made to contact Jemmy Button; there seems to have been an unspoken agreement (or possibly spoken, but certainly not written) that that chapter was closed. Setting out on the 9th, in the depths of southern winter, in spite of the inevitable bad weather the two ships stayed in company until 27 June, when the

Adventure suffered some damage to her main boom, and lagged behind. *Beagle* arrived at her old anchorage off Point Arena in Chiloé at midnight that day, and *Adventure* arrived two days later. At last the rigours of Tierra del Fuego were behind them. Resupplied and with minor repairs completed, they left for Valparaiso on 14 July, and arrived there together on the 22nd; but just at the time when thoughts of the optimists among them were turning to home and the warm waters of the central Pacific, FitzRoy received the news that provoked the most severe personal crisis of his life so far. The Admiralty would pay nothing towards the costs he had incurred in hiring the two small schooners and purchasing the *Adventure*, nor even contribute towards the costs of the extra seamen hired to man her.

FitzRoy had enough on his plate at this time without the arrival of this news from London (accompanied by an admonishment for having taken so long over the surveying on the east coast and in Tierra del Fuego!). There was so much work to do in drafting the charts based on the previous months of surveying (work described by FitzRoy as 'the dull routine of calculation', carried out in quarters rented ashore while the *Beagle* refitted) that he had not even taken time to visit the authorities in Santiago, sending Wickham, who spoke Spanish, in his place, to pay his respects and request the assistance of the Chileans in the next leg of their work. 'Nothing could be more satisfactory than the reply,' he wrote, 'and from that time until the Beagle left Chile she received every attention and assistance which the Chilean officers could afford.' Would that the Lords of the Admiralty had been so supportive, is the inference. The impact of the bad news is passed over by FitzRoy in a few words in his *Narrative*, where he refers only to a 'most painful struggle' in making up his mind what to do next. The sale of the *Adventure* was 'very ill-managed, partly owing to my being dispirited and careless', and raised only £1,400, slightly more than he had paid for her but much less than his total outlay including the refit and pay for her crew, and no reflection of her true worth; she was still trading on the coast of western South America, in good condition, when FitzRoy was writing up his *Narrative* in 1838. In a letter to Fanny written on 14 August, he makes light of the difficulties, but also gives a glimpse into his own mind:

Dearest Fanny do not let my spending money, in a good cause, disturb you. – I have ideas of money different from those of many persons – and I cannot see the wisdom of hoarding money with a view to the latter part of a life which is as precarious as the wind.

He was, after all, unmarried, with no dependants or responsibilities, and engaged on a hazardous occupation.

But although FitzRoy gives neither us nor his sister any real clues to his state of mind during this crisis, we have other accounts, notably from Darwin, to tell us just how bad things really were, although a private letter from FitzRoy to Beaufort, written between 26 and 28 September 1834, gives us some indication of his mental turmoil:

My schooner is *sold*. Our painting man Mr. Martens is *gone*. The Charts &c are progressing slowly – They are not ready to send away yet – I am in the dumps. It is heavy work – all work and no play – like your Office ... Troubles and difficulties harass and oppress me so much that I find it impossible either to say or do what I wish. Excuse me then I beg of you if my letters are at present short and unsatisfactory – My mind will soon be more at ease. Letters from my friends – Having been obliged to sell my Schooner, and crowd everything again on board the Beagle – Disappointment with respect to Mr. Stokes – also the acting Surgeon – and the acting Boatswain[16] – Continual hard work – and heavy expense – – These and many other things have made me ill and very unhappy.

Also on 28 September he writes to Fanny: 'My troubles are many and extremely oppressing – but I trust that time and exertion will overcome them.' But this time, his old remedy of hard work to take his mind off things was not enough.

Clearly, FitzRoy was suffering a nervous breakdown, brought on by overwork. Darwin, who had been away on one of his inland expeditions while the tedious work of drawing up charts was going on, returned just at this time, and was immediately made aware that something was wrong with the Captain. As he writes in his *Autobiography* (where, writing decades later, he misremembers the place as Concepción rather than Valparaiso):

> Poor FitzRoy was sadly overworked and in very low spirits; he complained bitterly to me that he must give a great party to all the inhabitants of the place. I remonstrated and said that I could see no such necessity on his part under the circumstances. He then burst out into a fury, declaring that I was the sort of man who would receive any favours and make no return. I got up and left the cabin without saying a word, and returned to Conception where I was then lodging. After a few days I came back to the ship and was received by the Captain as cordially as ever, for the storm had by that time quite blown over. The first Lieutenant, however, said to me: 'Confound you, philosopher, I wish you would not quarrel with the skipper; the day you left the ship I was dead-tired (the ship was refitting) and he kept me walking the deck till midnight abusing you all the time.'

Darwin himself was ill at this time – the only serious bout of illness he experienced during the voyage.[17] Some medical authorities suggest that he had become a victim of Chagas' disease (also known as South American sleeping sickness), a chronic condition resulting from a bite by a benchuca bug, and that this would explain his continuing illness in later life; whatever the cause of his illness, though, it meant that he saw little of FitzRoy during the next few weeks, when the Captain went through his personal crisis. But the *Beagle*'s officers, in particular Wickham, kept him informed. For some days in October 1834 FitzRoy withdrew into his cabin, and took no part in the running of the ship or the chart-making work. He consulted Bynoe, the Acting Surgeon, expressing his fear that he suffered from an inherited mental instability and was on the brink of going the same way as his uncle Castlereagh and his predecessor Stokes. And why should he carry on if the Lords of the Admiralty had no trust in his judgement? The only option, as he saw it, was to resign his commission on the grounds of ill health, hand the ship over to Wickham, and make his own way home. Looking thin and ill, he eventually emerged from the cabin with the letter of resignation written, and handed the ship over to his Lieutenant.

Or rather, he tried to. Wickham would have none of it, even though it would have meant a step up for himself. All the officers were doubly concerned at this development. First, they were concerned for their Captain, closing ranks to protect him from the consequences of taking an irrevocable step while not in his right mind. But in addition, the ship's orders were quite clear (as all such orders were) on what was to be done if the Captain died or was incapacitated. In this case, the First Lieutenant was required to bring the ship home by the shortest route – which would have meant going back to Tierra del Fuego and round the Horn. Such a prospect, after the past two winters and with the prospect of the warm waters of the Central Pacific beckoning, was just unthinkable. In a letter to his sister Catherine, written on 8 November, Darwin explains how the situation was resolved:

My last letter was rather a gloomy one, for I was not very well when I wrote it. Now everything is as bright as sunshine. I am quite well again, after being a second time in bed for a fortnight. Captain FitzRoy very generously has delayed the ship 10 days on my account & without at the time telling me for what reason. We have had some strange proceedings on board the *Beagle*, but which have ended most capitally for all hands. Capt. FitzRoy has for the last two months been working *extremely* hard, & at same time constantly annoyed by interruptions from officers of other ships: the selling the Schooner & its consequences were very vexatious; the cold manner the Admiralty (solely I believe because he is a Tory) have treated him, & a thousand other &c &c, has made him very thin & unwell. This was accompanied by a morbid depression of spirits, & loss of all decision & resolution. The Captain was afraid that his mind was becoming deranged (being aware of his hereditary predisposition), all that Bynoe could say, that it was merely the effect of bodily health & exhaustion after such application, would not do; he invalided & Wickham was appointed to the command. By the instructions Wickham could only finish the survey of the Southern part, & would then have been obliged to return direct to England. The grief on board the *Beagle* about the Captain's decision was universal and deeply felt. One great source of his annoyment was the feeling it impossible to fulfil the whole instructions; from his state of mind it never occurred to him that the very instructions order him to do as much of the West coast as *he has time* for, & then proceed across the Pacific. Wickham (very disinterestedly giving up his own promotion) urged this most strongly, stating that when he took the command nothing should induce him to go to Tierra del Fuego again; & then asked the Captain what would be gained by his resignation? Why not do the more useful part & return as commanded by the Pacific? The Captain at last to everyone's joy, consented & the resignation was withdrawn.

Hurra! Hurra! It is fixed the *Beagle* shall not go one mile South of Cape Tres Montes (about 200 miles south of Chiloe) & from that point to Valparaiso will be finished in about 5 months . . . For the first time since leaving England, I now see a clear & not so distant prospect of returning to you all; crossing the Pacific & from Sydney home will not take much time . . . When we are once at sea, I am sure the Captain will be all right again. He has already regained his cool inflexible manner, which he had quite lost.

FitzRoy may have been back to normal outwardly, but a letter to Fanny dated 6 November shows that he was only slowly recovering from the deep depression he had been in, although there is a heavy-handed attempt at humour: 'I am so surrounded with troubles and difficulties of every kind that I can only send a short and very stupid letter. My brains are more confused even than they used to be in London.'

Of course, the journey home took longer than anticipated. Although (having indeed waited for Darwin to recover from his illness), with all the charts to date completed and shipped off to England the *Beagle* sailed alone from Valparaiso on 10 November to recommence her surveying work, she would not leave South American waters, heading for the Galápagos Islands, until September 1835. As far as FitzRoy was concerned, most of the work was routine, and need not be described here. But there were two dramatic events in 1835 that were very far from routine.

The first was a major earthquake, the worst to hit the region for sixty or seventy years, accompanied by a tidal wave that affected parts of the coast, but had no severe effect on the *Beagle* itself. This struck on 20 February, when the ship was anchored in the port of Valdivia. FitzRoy and Darwin were both ashore at the time and felt the ground move, but Valdivia was only mildly affected by the tremor.[18] Darwin writes, 'I am afraid we shall hear of damage done at Concepcion . . . on board the motion was very perceptible; some below cried out that the ship must have tailed on the shore and was touching the bottom,' while FitzRoy reports that as they continued their surveying work over the next ten days:

> Shocks of earthquakes were frequently felt, more or less severely; sometimes I thought that the anchor had been accidentally let go, and the chain was running out; and while at anchor, I often fancied the ship was driving, till I saw that there was neither swell, current, nor wind sufficient to move her from the anchorage. We naturally concluded that some strange convulsion was working, and anxious for the fate of Concepcion, hastened to Talcahuano Bay as soon as our duty would allow: arriving there on the 4th of March – to our dismay – we saw ruins in every direction.

Both FitzRoy and Darwin give graphic accounts of the destruction; but the most salient point for us is the way they both observed evidence that the ground had been lifted up by the event, and the different ways in which they responded to the evidence of their own eyes.

FitzRoy was particularly struck by the changes visible around the coast of the Isla de Los Beyes, at the bottom of Concepción Bay: 'When walking on the shore, even at high-water, beds of dead muscles [mussels], numerous chitons and limpets, and withered sea-weed, still adhering, though lifeless, to the rocks on which they had lived, every where met the eye – proofs of the upheaval of the land.' But FitzRoy convinced himself that this was a temporary phenomenon, and that such sudden uplift events were followed by a gradual settling of the land back towards its former level:

> Whether this conjecture be well founded a short time may show: if it should be, an explanation might thus arise of the difference of opinion respecting the permanent elevation of land near Valparaiso, where some say it has been raised several feet in the last twenty years, while others deny that it has been raised at all. It may have been elevated, or upheaved as geologists say, for a time, but since then it may have settled or sunk down gradually to its old position.

Darwin, on the other hand, had by now ridden in the high Andes, and seen for himself rocks containing the fossil remains of marine creatures, such as those mussels that FitzRoy refers to, on the highest peaks. Together with the evidence of uplift that he had now seen for himself at sea level, this convinced Darwin that, very much in accordance with the ideas of Charles Lyell, the great bulk of the Andes had been raised to their present height by similar processes going on gradually over an enormous span of geological time. This 'gift of time' would later be essential in explaining how the slow process of evolution by natural selection could transform species into new forms; but it clearly flew in the face of the traditional interpretation of the Biblical chronology, even before Darwin turned his thoughts to the mechanism of evolution.

FitzRoy, however, had another explanation for the presence of fossilised shellfish on mountain tops:

> It appeared to me a convincing proof of the universality of the deluge. I am not ignorant that some have attributed this to other causes; but an unanswerable confutation of their subterfuge is this, that the various sorts of shells which compose these strata both in the plains and mountains, are the very same with those found in the bay and neighbouring places . . . these to me seem to preclude all manner of doubt that they were originally produced in that sea, from whence they were carried by the waters, and deposited in the places where they are now found.

This is part of an extended passage in the *Narrative* that was clearly written back in England, when FitzRoy was preparing his journal for publication, and not an on-the-spot comment made at the time. But it is significant as the first occasion on which he argues the literalist case, such as it is. By the time of writing he is aware of Darwin's interpretation of the evidence, having argued the case with him more than once, and refers angrily to this as a 'subterfuge'; but his 'confutation' of the argument is nothing of the sort, since even if all the fossil remains were the same as those of present-day inhabitants of the Pacific (which is not, in fact, the case), the explanation in terms of uplift would still be valid. Here, FitzRoy is bending the evidence and adapting his interpretation of the evidence to fit rigid, preconceived ideas. Although it is by no means clear that those ideas were already set in his mind as early as March or April 1835, the seeds of his eventual rift with Darwin were surely sown then.

But FitzRoy was far too busy surveying to worry much at the time about the age of the Earth and the interpretation of the Bible. Indeed, life could hardly have been better for him as the southern summer faded. Darwin was off on one of his extended trips into the interior while the *Beagle* was about her work, but on 22 April he was staying in Valparaiso when the ship hove to offshore, pausing only briefly on her way to send boats ashore. Apart from a natural desire to see his shipmates, Darwin had news from England which he couldn't wait to communicate to them – FitzRoy had been made Post.[19] In a letter to Susan Darwin dated 23 April, he writes: 'The Beagle passed this port yesterday. I hired a boat & pulled out to her. The Capt is very well; I was the first to communicate to him his promotion.' FitzRoy, although delighted, was, as ever, just as concerned about his subordinates. A short time earlier, he had expressed his views about his own long-delayed promotion in response to a letter from Fanny:

In your letter you speak about my not being yet promoted. – I wish Fanny I could get one or two of my hard working Shipmates promoted – it would gratify me much more than my own advance – which has been too tardy to be much valued – Plenty have gone over my head – one feels a *few* wounds but one gets hardened and

Robert FitzRoy was born on 5 July 1805 at the family home of Ampton Hall, near Bury St Edmunds, Suffolk. (Edifice © Gillian Darley)

FitzRoy's maternal uncle, Lord Castlereagh, was British foreign secretary from 1812 until his death by suicide in 1822. (Mary Evans Picture Library)

Robert FitzRoy in his early twenties. (Science Photo Library)

HMS *Beagle* in the Straits of Magellan in 1832. This was FitzRoy's second journey to South America and he carried on board the young Charles Darwin. (Mary Evans Picture Library)

A sketch of a Fuegian native at Portrait Cove, taken from FitzRoy's *Narrative*. (Art Archive)

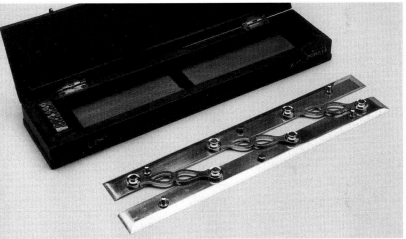

Top: Whaleboats like these (foreground) were invaluable on surveying voyages. The *Beagle* was unusually well equipped with them, carrying no less than four. (National Maritime Museum)

Centre: A set of parallel rules given to John Lort Stokes by FitzRoy and engraved with both their names. (National Maritime Museum)

Left: Darwin's sextant, used during his five-year voyage on the *Beagle*. (Bridgeman Art Library)

Charles Darwin circa 1840, shortly after settling at Down House in Kent. (Bridgeman Art Library)

One of Conrad Martens' famous sketches of the *Beagle*. Here the ship is being repaired on the shore of the Rio Santa Cruz in 1834. (Bridgeman Art Library)

John Lort Stokes' box, made from wood taken from the *Beagle* during her refit prior to the second South American voyage. (National Maritime Museum)

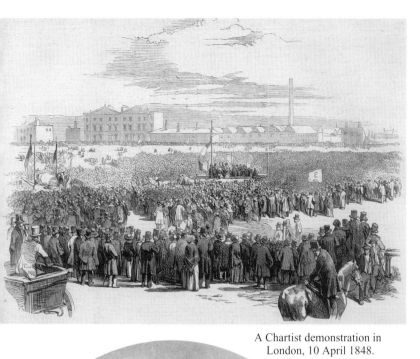

A Chartist demonstration in
London, 10 April 1848.
(Mary Evans Picture
Library)

The Crystal
Palace, home of
the Great
Exhibition of 1851.
(Science Museum/
Science and Society
Picture Library)

HMS *Arrogant* was FitzRoy's last command. She was the Navy's first screw-driven
steamship. (National Maritime Museum)

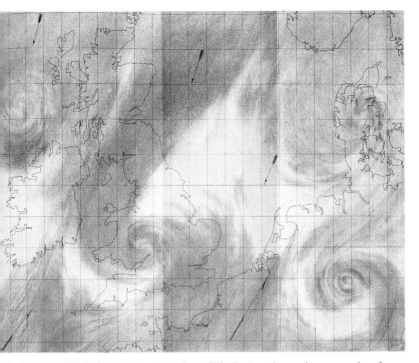

One of FitzRoy's early weather maps taken from his *Weather Book*. (Science Museum/Science and Society Picture Library)

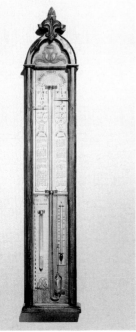

FitzRoy designed a reliable and simple barometer which quickly became standard issue. (National Maritime Museum)

Admiral Robert FitzRoy, FRS, photographed a short time before his death in 1865. (National Portrait Gallery)

FitzRoy's final resting place – Church of All Saints, south London. (Edifice © Philippa Lewis)

callous after receiving many. – Every one grumbles, you know, on this score, – & really I care as little as any one. I *deserve* it – for having burn't my fingers with politics – as they *tell* me. Truly though when I count the years I find that I have been *six* years a Commander – *Many* are only *one* year – But *many* more remain twice as long before they are passed on to the next shelf – it is a step of great consequence in the Navy because, afterwards, all goes by Seniority.

So when the news of his own promotion did come, FitzRoy's first reaction was typical:

I asked about Mr. Stokes and Lieut. Wickham, especially the former; but nothing had been heard of their exertions having obtained any satisfactory notice at head-quarters, which much diminished the gratification I might otherwise have felt on my own account. Mr. Darwin returned to the shore, intending to travel over-land, to meet us at Coquimbo, his very successful excursion across the Andes having encouraged him to make another long journey northward.

To add to his satisfaction with the way things were going, about this time the new Post Captain (still not quite thirty, and now assured of becoming an Admiral himself if he lived long enough) had found a way to speed up the work in spite of the difficulties placed in his way by the Lords of the Admiralty:

As another real benefactor to the public service, I may be allowed to mention Don Francisco Vascuñan, who lent me a vessel of thirty-five tons, called the Constitucion, to be employed in for-warding the survey. This craft was built in the River Maule, and bore a very high character as a sea boat. Lieutenant Sulivan, Mr. King, Mr. Stewart, and Mr. Forsyth volunteered to go in her . . . I despatched this new tender to examine a portion of coast near Coquimbo, which the Beagle had not seen sufficiently, and directed Lieut. Sulivan, if he found the vessel efficient, to continue afterwards surveying along the coast of Chile, as far as Paposo, whence he was to repair to Callao.

In high spirits at this success, FitzRoy soon had an opportunity to demonstrate all of his best characteristics as a naval officer. *Beagle* had returned to Valparaiso on 14 June to make preparations to leave Chilean waters and head northward. Loading of stores and provisions was well in hand when news came on 16 June of the wreck along the coast of HMS *Challenger* – the ship that had conveyed Lieutenant Smith to the Falklands, a three-masted vessel rather larger than the *Beagle*. Dispatches from Captain Michael Seymour, the commander of the *Challenger*, arrived the next day. The ship had been lost in foul weather on the night of 19 May, running ashore on rocks at the mouth of the Rio Leübu, well south of Concepción, when the Captain thought by his dead reckoning that they were well out to sea. Almost everyone from the ship was saved (with the aid of some local natives), and some help was provided by the British Consul in Concepción, but there was little that the shattered town could do to assist the shipwrecked sailors, who were waiting in a camp at the mouth of the Leübu.

FitzRoy's detailed account of the wreck and subsequent events played down his own part in saving the shipwrecked mariners; but Darwin (who was still away on an inland expedition at the time, but got first-hand reports on his return to the ship) and the officers of the *Beagle* were less reticent in their accounts. The senior British officer in Valparaiso was Commodore Mason, in the frigate HMS *Blonde*. FitzRoy tells us that as the *Blonde* prepared for sea to go to the aid of the *Challenger*'s men:

An offer of such assistance as I could render was accepted by the commodore; and, having arranged the Beagle's affairs, as far as then necessary, I went on board the Blonde, taking with me Mr. Usborne, J. Bennett, and a whaleboat. Lieut. Wickham was to forward the Beagle's duty during my absence, and take her to Copiapo [where Darwin would meet the ship], Iquique, and Callao, before I should rejoin her.

It was natural that the Commodore should take advantage of FitzRoy's detailed knowledge of the coastal waters; but there is more to this story, as Darwin reveals in a letter to Caroline Darwin simply dated July 1835: 'The old commodore of the Blonde was very slack in his motions – in short, afraid of getting on that lee-shore in the winter; so that Capt FitzRoy had to bully him and at last offered to go as Pilot.' It seems that a certain amount of hot coffee had to be poured over the senior officer (by one of the most junior Captains on the Navy List) before he was willing to put to sea at all and FitzRoy's 'offer' was accepted! But it is perhaps worth mentioning that Mason already knew FitzRoy – back in November 1834 he had written to the Hydrographer from Valparaiso praising FitzRoy's work, and mentioning that he had been ill but was now getting better. Mason may have been over-cautious, but when push came to shove, he knew that FitzRoy was a better sailor than he was.

On 21 June the Blonde anchored at Talcahiano, the nearest port to the site of the shipwreck, where Commodore Mason's nervousness was further increased when the Port Captain told him that in his opinion the encampment at the mouth of the Leübu was 'quite inaccessible in any weather' by a ship, but that boats could get in to the mouth of the river. One of the Challenger's officers, Lieutenant Collins, was at the port, and he told FitzRoy that his shipmates were running low on supplies, that they were becoming sickly, and that they were threatened by a large party of hostile Indians. FitzRoy, preferring to trust his own judgement rather than that of the Port Captain, decided to go there on horseback to check out the situation for himself, with 'orders and letters from Commodore Mason'. We wonder who actually drafted those orders. After a wild ride across country, FitzRoy and a small party reached the encampment on the 23rd. On the 26th, he was back at the Blonde with his assessment of the situation and a letter from Captain Seymour to Commodore Mason.

While FitzRoy had been away, the Commodore (apparently in an attempt to shift the responsibility for anything remotely dangerous on to other shoulders) had hired an American schooner to go to the aid of Captain Seymour's party, and sent with this vessel FitzRoy's men and his whaleboat. This proved a complete waste of time – the schooner turned out to be in such a poor state that it played no part in the rescue and a few days later (after the rescue) it had to be taken in tow by the *Blonde* and rescued itself. Fortunately, with FitzRoy back the Commodore was bullied, or shamed, into leaving port on the 27th. But first severe weather and a strong wind kept them out of sight of shore while heading south to the Leübu, then thick haze prevented them from identifying their landmarks – 'scarcely indeed could we discern the line of the surf, heavily as it was beating upon the shore; and at noon we were obliged to haul off, on account of wind and rain'. When the weather did clear on 5 July, after 'the longest week I ever passed', the *Blonde* was becalmed almost five miles from land. The smoke and flags of the camp were in sight from early morning, yet to FitzRoy's frustration 'no steps were taken until near one o'clock, though it was a beautiful, and almost calm day'.

Eventually, FitzRoy got permission to take a party in boats ashore. There, he found the *Challenger*'s crew in deteriorating health, but intact. It was too late to attempt to take them off by boat that evening, but with men from the *Challenger* added to the crew of the barge to double-bank the oars, 'I hastened out of the river as the sun was setting. A light breeze from the land favoured us, and though the Blonde was hull down in the south-west when we started, we were happy enough to get on board at about eight o'clock,' in the midwinter darkness.

On Mason's orders, as soon as the barge was hoisted in 'the frigate again made sail off shore', to avoid any risk of running aground. 'But,' FitzRoy tells us disingenuously, while Mason was asleep below 'a fortunate mistake caused the main-yard to be squared about midnight, and at daybreak next morning we were in a good position off the entrance of the river'. Clearly, the 'fortunate mistake' was in the same category as Nelson's failure to see the signal when he clapped a telescope to his blind eye. The rescue was accomplished without further incident (except for the recovery of the dismasted American schooner, which would probably have been lost entirely if it had not been for the efforts of the party from the *Beagle*). The officers and crew of the *Challenger* were in no doubt who they had to thank. Their accounts of the shipwreck include many references along the lines of 'Captain Fitzroy, to whom we owe a debt of great gratitude for his zealous exertions on our behalf', and 'our tried friend Captain Fitzroy'. (Similar remarks about Commodore Mason are conspicuous by their absence.) FitzRoy was able to do them one more favour. At the inevitable court martial (the commander of a Royal Navy ship is always court martialled if the ship is lost) Seymour produced in his defence a paper from FitzRoy (now in the Public Record Office) suggesting that the ocean currents in the region had been altered by the earthquake (which Seymour, having just come round the Horn, knew nothing of at the time) and that this was the reason for the error in dead reckoning which caused the wreck. Seymour was exonerated, praised for his conduct following the shipwreck, and given another ship.

There was nearly another court martial in connection with the affair. In his letter of late July, on board the *Beagle* and awaiting the return of the Captain, Darwin continued:

We hear that they have succeeded in saving nearly all hands, but that the Captain & Commodore have had a tremendous quarrel; the former having hinted something about a Court-Martial to the old Commodore for his slowness. – We suspect that such a taught hand as the Captain is, has opened the eyes of everyone fore & aft in the Blonde, to a most surprising degree. We expect the Blonde will arrive here in a very few days and all are very anxious to hear the news; no change in state politicks ever caused in its circle more conversation than this wonderful quarrel between the Captain & the Commodore has with us.

Darwin had rejoined the *Beagle* at Copiapò on 5 July, the very day that the *Blonde* made contact with the shipwrecked mariners, so news must have travelled fast up the coast. But in spite of his hopes FitzRoy did not join them until 9 August, in Callao Bay, as he had originally intended when leaving Wickham to 'forward the Beagle's duty'. The rendezvous had been chosen because that was the place where Sulivan was to meet up with the *Beagle* again after completing his independent surveying work in the *Constitución*, and he duly arrived on 30 August. Sulivan's cruise had been so successful, and he gave such a glowing account of the quality of the little vessel, that FitzRoy, fully restored to his old self after the events of the previous few weeks, seized the opportunity to get home quickly while both having his cake (completing the survey work) and eating it as well (completing the meridian chain). You can guess – he bought the *Constitución* for £400 and, since Sulivan could not be spared, left Alexander Usborne, the Assistant Sailing Master of the *Beagle*, in command of her, with Midshipman Charles Forsyth and a Master's Assistant from the *Blonde*, Mr E. Davis, who 'Commodore Mason was prevailed upon to allow' to join the little expedition (by now, Mason would probably have done anything to speed FitzRoy on his way). 'Seven good seamen, and a boy, volunteers from the Beagle, completed Mr. Usborne's party.' Their orders were to spend eight or ten months completing the survey of the whole coast of Peru in their little ship (all of 35 tons, remember, compared with the *Beagle*'s 242 tons), then to sell the vessel and seek the assistance of the British Consul in Peru in obtaining passage back to England on a vessel with 'the necessary accommodation which you will require in order to prosecute your work [on the charts] during the homeward passage' back around the Horn to England. It goes without saying that they completed the task on time and with utter aplomb; they had, after all, been trained by Robert FitzRoy.[20]

The *Constitución* sailed on 6 September 1835, and the *Beagle* followed her out of harbour, homeward bound at last and with everyone on board in high spirits, on the following day. In his last letter to Fanny from South America, dated 7 September, FitzRoy proclaims the good news: 'This day the Beagle leaves the Coast of South America . . . In March she will be in Sydney in N. S. Wales – in July at the Cape of Good Hope and in October 1836 – in old England – Glorious Prospect.'

For the meridian observations, the route included stops (with a little light surveying) at the Galápagos Islands, Tahiti, New Zealand, Australia (where Midshipman King left to rejoin his father, having decided that the Navy life was not for him), the Keeling Islands, Mauritius, the Cape of Good Hope, and several islands in the Atlantic. To the frustration of many of those on board, FitzRoy also decided to touch again at Bahia, on the coast of South America, to recheck his observations and complete the circumnavigation of the globe, before heading north to England. But we can skip over most of the events of this leg of the voyage, which was largely routine as far as FitzRoy was concerned, although the visit to the Galápagos Islands in particular played a famously significant role in the development of Darwin's ideas.

Two of the ports of call on the voyage home were to be the scenes of events with a particular resonance for FitzRoy, with his interest in the uplift of primitive people with the aid of Christian teaching. On Tahiti, he found what seemed to be an almost ideal example of the success of such missionary work. The *Beagle* arrived there on Sunday 15 November by the ship's log; but on the island it was already Monday 16 November. The ship had 'lost' a day by having sailed so far westward, and on the following day FitzRoy noted that it was 'with us the 16th, but to agree with the reckoning of Otaheite [Tahiti] and those who came from the west, changed to the 17th.' So for the *Beagle* and her crew, there was no 16 November 1835. The natives were friendly, almost teetotal (refusing spirits entirely but accepting a little wine when entertained on board the ship) and strictly observed the Sabbath. The most disreputable person FitzRoy met there, indeed, was a European adventurer – the self-styled Baron Charles Philippe Hippolytus de Thierry, who had been born in England of French émigré parents in 1795. After a series of escapades, he had obtained some sort of title to 40,000 acres of land in New Zealand, where he was planning to travel to set himself up as King, but had stopped over in Tahiti. He was surrounded by

> a motley group of tattowed New Zealanders, half-clothed natives of Otaheite, and some ill-looking American seamen ... In his house was a pile of muskets, whose fixed and very long bayonets had not a philanthropic aspect. He had been there three months, and was said to be waiting for his ships to arrive and carry him to his sovereignty ... he had succeeded in duping a great many people.

FitzRoy was more impressed by the court of the native queen, Pomare. He had what could have been unpleasant business to conduct with her, a duty imposed on him by Commodore Mason. Back in 1831, the Master and Mate of a British ship, the *Truro*, had been murdered by natives of islands to the east of Tahiti which also came under the queen's jurisdiction. In recompense, Captain Seymour, on a visit in HMS *Challenger*, had extracted a promise that a fine of 2,853 dollars would be paid by 1 September 1835. The fine not having been paid, FitzRoy was instructed to take the necessary steps. The case was made by FitzRoy, accompanied by all the officers that could be spared from their duties with the ship, through an interpreter at a formal gathering of chiefs presided over by the queen:

> Disposed at first to criticise rather ill-naturedly – how soon our feelings altered, as we remarked the superior appearance and indications of intellectual ability shown by the chieftains, and by very many of the natives of a lower class. Their manner, and animated though quiet tone of speaking, assisted the good sense and apparent honesty of the principal men in elevating our ideas of their talents, and of their wish to act correctly.

Everything went smoothly. The fine was paid, and before the *Beagle* departed they gave a dinner for the queen and her entourage, with a very well-received display of signal rockets afterwards.

> I much wished then to have had a few good fireworks of a more artificial character. To any one about to visit distant, especially half-civilized, or savage nations, let me repeat a piece of advice given to me, but which from inadvertence I neglected to follow: 'Take a large stock of fireworks.'

They sailed on 26 November, with both FitzRoy and Darwin having been impressed by the success of the missionaries in 'civilising' the Tahitians. At their next port of call, they would see an example of the raw material which might benefit, in their eyes, from such a civilising influence.

Although the *Beagle* made out the northern hills of New Zealand on 19 December, adverse winds prevented the ship from reaching Kororareka harbour, in the Bay of Islands, until the 21st. This was the site of the principal settlement at that time, and they were at first surprised by the lack of a proper welcome – only one boat put out from the shore, approaching the ship cautiously. It turned out that they had been suspected of being a private vessel carrying the so-called Baron de Thierry, partly because of the number of boats they carried on deck. Once the truth was known, 'visitors hastened on board. Had he made such an experiment, he would hardly have escaped with his life, so inveterate and general was the feeling then existing against his sinister and absurd attempt. He would indeed have found himself in a nest of hornets.'

There was no government in New Zealand, just a single official 'British resident', James Busby, with a house and a tall flagstaff on which waved the Union Flag.[21] Neither FitzRoy nor Darwin was impressed with Kororareka; the latter wrote in his diary:

> This is the largest village & will one day no doubt increase into the chief town. Besides a considerable native population there are many English residents. – These latter are of the most worthless character; & amongst them are many run away convicts from New South Wales. There are many spirit shops, & the whole population is addicted to drunkenness & all kinds of vice. As this is the capital, a person would be inclined to form his opinion of the New Zealanders from what he here saw; but in this case his estimate of their character would be too low. – This little village is the very strong-hold of vice; although many tribes, in other parts, have embraced Christianity, here the greater part are yet remain in Heathenism.

It was, in short, a frontier town of the worst kind.

Things were much better on the pretty island of Pahia, in the Bay, where the missionaries and the British resident lived, and where FitzRoy landed his instruments to make his scientific observations. It was there, also, that they attended divine service on Christmas Day, their fourth since leaving England and fifth, including the one spent in Plymouth, since the ship had been commissioned. On a visit to Waimate, about 15 miles inland from the Bay of Islands, Darwin saw the potential of New Zealand – 'the sudden appearance of an English farm house & its well dressed fields, placed there as if by an enchanter's wand'. He took tea with the family, and was shown around the farm:

> Fine crops of barley & wheat in full ear, & others of potatoes & of clover were standing; but I cannot attempt to describe all I saw; there were large gardens, with every fruit & vegetable which England produces & many belonging to a warmer clime. – I may instance asparagus, kidney beans, cucumbers, rhubarb, apples & pears, figs, peaches, apricots, grapes, olives, gooseberries, currants, hops, gorse for fences, & English oaks!

There was also livestock, and buildings including a substantial watermill:

> Native workmanship taught by the Missionaries has effected this change ... When I looked at this whole scene I thought it admirable ... Nor was it the triumphant feeling at seeing what Englishmen could effect: but a thing of far more consequence; – the object for which this labor had been bestowed, – the moral effect on the native inhabitant of New Zealand.

FitzRoy also saw both the present difficulties and the scope for moral uplift in New Zealand. On the one hand, he was pressed into acting as a kind of unofficial magistrate to settle various disputes (a duty he described as involving 'unpleasant discussions' on 'a maze of disagreeable questions, in which it was not my proper business to interfere'), since there was no law of any kind and the British resident had no duties or responsibilities except literally to fly the flag. On the other hand, FitzRoy had conversations with the missionaries, which clearly made a deep impression on him and like Darwin he saw the success of their efforts at introducing English-style farms in New Zealand:

> It was also very gratifying to me to mark the lively interest taken by [the missionaries] Mr. Williams, Mr. Davis, and Mr. Baker in every detail connected with the Fuegians, and in our attempt to

establish Richard Matthews in Tierra del Fuego. Again and again they recurred to the subject, and asked for more information; they could not hear of my calling the attempt 'a failure.' 'It was the first step,' said they, 'and similar in its result to our first step in New Zealand. We failed at first; but, by God's blessing upon human exertions, we have at last succeeded far beyond our anticipations.' Their anxiety about the South American aborigines generally; about the places where missionaries might have a chance of doing good; and about the state of the islands in the Pacific Ocean, gave me a distinct idea of the prevalence of true missionary spirit.

One happy outcome of this visit was that Matthews decided to stay in New Zealand, where his brother, also a missionary, was established and had recently married the daughter of Mr Davis. Without Matthews, but with plenty of food for thought, FitzRoy took the *Beagle* out of the Bay of Islands on 30 December. The strain of the past four years had left its mark, and after the ship called at Sydney, on 2 February 1836 Philip King wrote in a letter to Beaufort that 'I regret to say he has suffered very much and is yet suffering much from ill health – he has had a very severe shake to his constitution which a little *rest* in England will I hope restore for he is an excellent fellow and will I am satisfied yet be a shining ornament to our service.'

There was very little prospect of much immediate rest when the *Beagle* reached England, with all the paperwork connected with the voyage to complete; but by FitzRoy's standards the rest of the voyage home was almost a holiday. In a letter to Caroline Darwin, written from Mauritius and dated 29 April 1836, Darwin says that:

The Captain is daily becoming a happier man, he now looks forward with cheerfulness to the work which is before him. He, like myself, is busy all day in writing, but instead of geology, it is the account of the Voyage. I sometimes fear his 'Book' will be rather diffuse, but in most other respects it certainly will be good: his style is very simple & excellent. He has proposed to me, to join him in publishing the account, that is, for him to have the disposal & arranging of my journal & to mingle it with his own. Of course

I have said I am perfectly willing, if he wants materials; or thinks the chit-chat details of my journal are any ways worth publishing. He has read over the part I have on board, & likes it.

With the detour to Bahia, the *Beagle* eventually anchored at Falmouth on 2 October 1836, after an absence of four years and nine months from England. FitzRoy's first action was to pen a letter to Fanny:

Returned! Every one well – and the little ship uninjured.

How are you? – where are you? – where are you likely to be during this month? How are your dear and constant little companions?

. . . Our voyage has been more successful than I had any right to anticipate. We have been *extremely* fortunate in *all* ways. Do not be vexed at my saying that the results of it will require two years application, in good earnest, – in London . . .

My Dearest Sister I pray that you are as well and as happy as I am – and anxiously wait to hear – direct to me at Devonport. Your most affectionate Brother Robert.

At Falmouth, as we have seen, Darwin left the ship, hurrying home overland to Shrewsbury. But FitzRoy's voyage was not yet quite complete.

CHAPTER SIX

The Happy Return

AFTER DROPPING DARWIN at Falmouth, the *Beagle* then sailed to Greenwich, via Plymouth, Portsmouth and Deal, dropping anchor there on 28 October 1836 for FitzRoy to carry out the final set of chronometric measurements for the meridian chain. If his chronometers had kept perfect time, they should by now have been exactly twenty-four hours adrift from Greenwich time; in fact, they showed a discrepancy from the expected time, after nearly five years at sea, of just thirty-three seconds. FitzRoy had put a chronometric girdle around the Earth with almost exquisite precision, establishing a basis for precise navigation that would alone have been ample justification for the voyage. Had he really needed all those chronometers that he had been at such pains to acquire at the start of the voyage? In the Appendix to the *Narrative*, he sums up their fates:

Some of the watches stopped; others altered their rates suddenly; and in one case (R) a mainspring broke when the chronometer had been going admirably, till that moment. Four chronometers were left with Mr. Usborne, on the coast of Peru, and in consequence of these diminutions of our original number, there were but eleven watches in tolerably effective condition during the last two principal links of the chain, namely, from Port Praya to the Azores, and from the Azores to Devonport.

In other words, only half the original number completed the voyage in working order; FitzRoy's caution had been well justified.

In the perfect ending to the voyage, while they were at Greenwich Usborne and his party, freshly returned from their independent survey of the coastline of Peru, rejoined the ship. There were other visitors, and one particular incident provided an anecdote for Sulivan's biography:

On the return of the vessel after such an interesting voyage, so many people came to visit her that the captain gave the order that *respectable-looking persons only* were to be admitted by the accommodation-ladder; others were to enter by the gangway (where some projections three inches wide against the ship's side afford foothold, there being two ropes to assist the climber). Sulivan, who was at the time on watch, noticed the sentry wave a boat away from the ladder round to the gangway. Presently the head of a very pretty, stylish woman appeared in it, and Sulivan went forward to assist her. She was followed by a rather plain-looking man, who asked for the captain. After they had been conducted below, FitzRoy came on deck, much put out, and said, 'Do you know it was the Astronomer Royal who has been treated with such scant ceremony?' He was paying what was somewhat of an official visit, with his wife. When the captain had retired below, Sulivan rated the sentry for his want of discrimination. The man replied, 'Well, sir, he did not *look* respectable!'

It is unlikely that the Astronomer Royal, George Airy, would have been much troubled by the sentry's mistake; he had only taken up his post in 1835, finding the Royal Greenwich Observatory in a shambolic state but transforming it during his long reign (he held the post until 1881) into the finest observatory of its kind in the world, through the same kind of ruthless efficiency that FitzRoy applied in his own field. He would have been much more interested in the chronometric observations than in ceremonial. It's a sign of just how much interest there was in the voyage that FitzRoy read a paper about the surveying work and their experiences of earthquake activity to the Royal Geographical Society while the ship was still in Greenwich, and before it was finally paid off at the end of the voyage.[1] As he told his audience, 'beginning with the right or southern bank of the wide river Plata, every mile of the coast thence to Cape Horn was closely surveyed', as well as the Falklands, and although westward of Cape Horn there had been little need to add to the work carried out by the *Beagle* on her previous voyage up to latitude 47° south, 'between forty-seven south latitude and the River Guayaquil, the whole coasts of Chile and Peru have been surveyed; no port or roadstead has been omitted'. And, of course, copies of the relevant charts were given to the governments of the various countries where the surveys were carried out. But the talk also contains just a hint of the pressure FitzRoy had been working under for the previous five years. After giving handsome credit to the burden carried by his junior officers when it was impossible for him to be in two places at once, he remarks, 'here let me quietly protest against the attempted union of petty astronomer, experiment maker, and captain of a man-of-war in one individual'.

But his work in connection with the voyage was far from over. The very last leg of the epic voyage was from Greenwich along the River Thames to Woolwich, on 7 November, where the *Beagle* was paid off on 17 November 1836, just one month short of five full years since she had left Plymouth, and five years and five months (to be precise, five years and 136 days) since she had been commissioned on 4 July 1831. In all that time, just three members of the crew had died of the fever contracted while ashore in Rio, one in the duck-shooting accident, and the purser of natural causes – essentially, old age:

no serious illness, brought on or contracted while on service, happened on board; neither did any accident of consequence occur in the ship; nor did any man ever fall overboard during all that time . . . our immunity from accident during exposure to a variety of risks, especially in boats, I attribute, referring to visible causes, to the care, attention, and vigilance of the excellent officers whose able assistance was not valued by me more than their sincere friendship.

While FitzRoy rightly, and characteristically, emphasises the human side of the *Beagle*'s safety record, Sulivan tells us that during the entire voyage they did not lose a single spar – an equally eloquent tribute to FitzRoy's care in fitting out the ship and skill at handling her. In an appreciation published after his death,[2] an anonymous fellow officer (probably Stokes or Sulivan) comments:

When it is recollected that the *Beagle* was one of the so-called 'coffin' 10-gun brigs, that she always went to the south for her work, so loaded with provisions that her copper was under water, and that both upper deck and poop were stowed thickly with salt meat casks, and that she had six large boats for her class – two over head on studs – it will give an idea of the care and seamanship required to produce such results.

The same testimonial mentions that:

An officer who served with Admiral FitzRoy for the first time in the *Beagle* in 1831, and was a mate [that is, passed Lieutenant] of four years' standing when he joined her,[3] once remarked: 'If any one had told me that I was not a seaman when I joined this ship, I should have been greatly offended, but now I know that I never knew what real seamanship was until I saw it in this vessel.'

Of the 'steady support and unvarying help' which he received from those officers, FitzRoy says that 'where all did so much, and all contributed so materially to the gatherings of the voyage, it is unnecessary to particularise, farther than by saying that Mr. Stokes's services hold the first place in my own estimation'. Which makes Stokes, who was the Assistant Surveyor on the voyage, but only a Mate who had not yet received his Lieutenant's commission, an outstanding officer indeed. David Stanbury has listed some of the notable positions held by the officers from the *Beagle* (including FitzRoy and Darwin) in later life:

> No less than five of the *Beagle*'s officers were destined to reach the rank of admiral; two became captains of the *Beagle*; two, eventual Fellows of the Royal Society. They also included Governor Generals-to-be of New Zealand and Queensland; a Member of Parliament; future Heads of the Board of Trade and the Meteorological Office; two artists who achieved considerable renown in the country of their adoption; three doctors; the prospective Secretaries of the Geological Society and the Royal Geographical Society; an Inspector of Coastguards; an Australian property magnate; the founding father of the British colony of the Falkland Islands; six highly professional surveyors; four botanists of sufficient standing to correspond with the great Hooker at Kew; five active collectors whose specimens were to be eagerly described by the Zoological Society and the Natural History Museum; one of the founders of the science of meteorology; and the author of *The Origin of Species*. By any standards a talented bunch!

It seems genuinely never to have occurred to FitzRoy (although it did not escape the notice of people like Beaufort) that this remarkable list of achievements from such a small number of men must have owed at least something to their serving for so long under his command – as we have noted, even Darwin learned the meticulous methods of work which were to characterise his later life from the Captain.

The extent to which this was, indeed, a happy return can be judged from an exchange of letters which took place between Darwin and FitzRoy even before the *Beagle* had been paid off. Remember that Darwin was now twenty-seven, and had spent almost one-fifth of his entire life on the voyage; FitzRoy was now thirty-one, and his world had revolved around the *Beagle* for eight of those thirty-one years — more than a quarter of his life. Their reaction now was like schoolboys escaping from their cloistered world at the end of a long, hard term. On 6 October 1836, Darwin wrote to 'My dear FitzRoy' from Shrewsbury:

I arrived here yesterday morning at Breakfast time, and, thank God, found all my dear good sisters & father quite well. My father appears more cheerful and very little older than when I left. My sisters assure me I do not look the least different, & I am able to return the compliment. Indeed, all England appears changed except the good old Town of Shrewsbury & its inhabitants, which for all I can see to the contrary, may go on as they now are to Doomsday. I wish with all my heart I was writing to you amongst your friends instead of at that horrid Plymouth. But the day will soon come, and you will be as happy as I now am. I do assure you I am a very great man at home – the five years' voyage has certainly raised me a hundred percent. I fear such greatness must experience a fall.

I am thoroughly ashamed of myself in what a dead and half alive state I spent the few last days on board; my only excuse is that certainly I was not quite well. The first day in the mail tired me, but as I drew nearer to Shrewsbury everything looked more beautiful & cheerful. In passing Gloucestershire and Worcestershire I wished much for you to admire the fields woods & orchards. The stupid people on the coach did not seem to think the fields one bit greener than usual, but I am sure we should have thoroughly agreed that the wide world does not contain so happy a prospect as the rich cultivated land of England.

I hope you will not forget to send me a note telling me how you go on. I do indeed hope all your vexations and trouble with respect to our voyage, which we now *know* has an end, have come to a close. If you do not receive much satisfaction for all the mental and bodily energy you have expended in His Majesty's service, you will be most hardly treated. I put my radical sisters into an uproar at some of the *prudent* (if they were not *honest* Whigs, I *would* say shabby) proceedings of our

Government. By the way, I must tell you for the honor & glory of the family, that my father has a large engraving of King George the IV, put up in his sitting room. But I am no renegade, and by the time we meet, my politics will be as firmly fixed and as wisely founded as ever they were.

I thought when I began this letter I would convince you what a steady & sober frame of mind I was in. But I find I am writing most precious nonsense. Two or three of our labourers yesterday immediately set to work, and got most excessively drunk in honour of the arrival of 'Master Charles'. Who then shall gainsay if Master Charles himself chooses to make himself a fool.

Goodbye – God bless you – I hope you are as happy, but much wiser than your most sincere but unworthy Philos. Chas Darwin

FitzRoy replied to 'Dearest Philos' from Portsmouth on 10 October:

What will you say to me for not having written before I know not – but really I have *not* been idle or forgetful.

I trusted to Fuller[4] for all immediate necessary information and I will write now to give you the rest.

Captain Beaufort was out of town when my letter & papers reached London (from Falmouth) and the Chart duster put them *away* in a *corner* (excepting *one* private one) to await Capt. B's *return*!!! Those papers related to the chronometric results &c. &c. – upon which the necessity of our going to Woolwich was to be founded – orders had been sent to Plymouth for the *Beagle* to pay off there – but L^d. Amelius Beauclerk had civility and sense enough to stay proceedings and approve of my going to London to see the Lords & Masters myself – I boarded Sir John Barrow,[5] and then made a stalking horse of *him* while attacking the *others*.

All was satisfactorily settled in a very short time – and they acceded civilly to my proposals of calling at *Portsmouth*. I was delighted to see that the Valp^o. cargo of charts had not only *arrived* but they were mostly *Engraved* – or in the Engraver's hands – and on a *large* scale! They have given much satisfaction at the Hyd^l. Office.

I have promised to give them a short paper for the Geog[l]. Society – a slight *sketch* of our voyage – I will do what I can – according to time and ask you to add and correct.

Rice Trevor & Alex[r]. Wood[6] crossed me on the road – they in one mail – I in another – but I was soon down again & with them at Devonport – Fuller told me you looked very well and had on a *good hat*!

Who the deuce was *my cousin* in a *broad brimmed* hat?

I was delighted by your letter, the account of your family – & the joy tipsy style of the whole letter were *very* pleasing. Indeed, Charles Darwin, I have *also* been *very* happy – even at that horrid place Plymouth – for that horrid place contains a *treasure* to *me* which even you were ignorant of!! Now guess – and think & guess again. Believe it, or not, – the news is *true* – I am going to be *married*!!!!!! To Mary O'Brien.

Now you may know that I had decided on this step, long *very long* ago. All is settled & we shall be married in December. Rice Trevor, Alex[r]. Wood & Talbot like her *much*. Pray call on my sister in Stratton Street – she longs to see you, – and ask to see the *children*.

Money matters are better than *you* think. Your's most sincerely Rob[t]. FitzRoy

The news of FitzRoy's marriage (which took place on 27 December 1836) comes as much as a bolt out of the blue to the historian as it clearly did to Darwin; indeed, we know very little about FitzRoy's family life as a married man. Although he would in fact marry twice, and had five children who survived infancy, his wives and daughters remain shadowy figures, and only his son, who went on to have a distinguished career in the Navy and therefore left a mark in the official records, emerges from the shadows (see Appendix I). Whereas the Darwin family seem to have been obsessive hoarders of letters (especially after Darwin's fame began to grow), there is no surviving comparable source of information about FitzRoy and his family. But we do at least know that Mary Henrietta O'Brien was the daughter of a retired army officer and country gentleman, a widower, Major General

Edward James O'Brien, who had been born in the first half of the 1780s. The fact is, though, that FitzRoy's family life would always come second to his work, except for a period of domestic tragedy in 1852, so that the absence of much in the way of domestic detail is of no great significance in trying to understand what made him tick – or rather, the absence of domestic detail is an accurate reflection of the priorities in FitzRoy's life. For the rest of his days, he remained driven by that sense of duty and obligation to act for the public good, whatever the personal consequences, which shows up so clearly in his period in command of the *Beagle*.

FitzRoy's duty in the years immediately following the return of the *Beagle* to England was clear. His first priority was to complete the preparation of the mass of charts, sailing directions, and other technical material resulting from the voyage, and he continued to receive his pay for carrying out this work even after the *Beagle* was paid off. We shall not go into any of the details, except to note that work on this material continued long after the pay for it stopped, and that the last instalment was sent back to Beaufort by FitzRoy on his arrival in New Zealand in September 1844, having presumably been completed on the voyage out from England. But there was also considerable public interest in the activities of the *Beagle*, and FitzRoy also felt it his duty to write up the material from both voyages into a book. With his predecessor, Pringle Stokes, dead, and with Captain King having retired to Australia, FitzRoy carried the responsibility (as he saw it) for all the material from the first voyage, as well as his own narrative of the second voyage. It soon became clear that with all this, and Darwin's material as well, there was far too much for one volume. We do not have the details of the discussions that must have taken place, but it is clear that Darwin and FitzRoy soon reached an amicable agreement (by January 1837) that FitzRoy would use the material he was responsible for to prepare two volumes, one covering the first voyage (*Beagle* and *Adventure*) and the other the second, while Darwin would write his own account of the second voyage for a third volume to be published with the other two as a single book.[7]

As the work progressed, FitzRoy found that the only way to cope with the mass of technical material and his own observations on various topics (such as the Fuegian language) that would otherwise break up the flow of the narrative was to relegate the material to a fourth volume of Appendices. Altogether he would produce well over half a million words, all written out by hand. It is small wonder that in a letter to Beaufort dated 13 December 1837, when the project was still far from completion, he would write: 'I am extremely like an ass between two bundles of hay – The Charts and the book are each hailing me and tell me they require my undivided attention – to do them full justice.'

By the time the book actually appeared, two years later, FitzRoy was no longer, at his own request, being paid to work on 'The Charts'. On 3 January 1839, he had written to Beaufort:

As the time for Estimates approaches I send you a formal application to be removed from your list of people receiving extra pay; which will lighten your estimates for the ensuing year. But by this request I do not imply the slightest intention of withdrawing from your neighbourhood – or leaving any part of the survey in an unfinished state: I shall go on with the work I have yet to do – just as usual.

Two years is a long time for any man to receive extra pay for completing his surveys, – and most of that time has been occupied in my *Narrative*, – which cannot be considered a Government Work.

This is all utterly characteristic. In spite of the shabby way that he has been treated by the Admiralty concerning his expenditure on auxiliary vessels, leaving him at least £3,000 out of pocket, FitzRoy is anxious to do the honourable thing and save Beaufort's office the few pounds represented by his own pay for the coming year. But he will see that the work is completed – at his own expense! Why? For the public good, of course.

FitzRoy's lack of financial nous – or genuine unconcern about money – led him to make a deal for the publication of the book which would later rankle with Darwin. The publisher, Henry Colburn, should have paid the authors a flat fee for the book, which eventually appeared in May 1839, but it seems that any cash flow was in the other direction.

Nora Barlow[8] quotes a letter from Darwin to his sister Susan in which he says: 'talking of money, I reaped the other day all the profit which I shall ever get from my *Journal*, which consisted in my paying Mr. Colburn £21 10s. for the copies which I presented to different people; 1,337 copies have been sold. – This is a comfortable arrangement, is it not?' In 1845, Darwin would actually receive £150 when he sold the copyright in the second edition of his book to Murray; but FitzRoy never made anything out of his *Narrative*.[9] Nevertheless, this was a time of double celebration for FitzRoy, whose son and second child Robert O'Brien FitzRoy had been born on 2 April, just a month before the book was published. By popular demand, Darwin's volume alone was reprinted the same year (no doubt mildly to the chagrin of FitzRoy), with no payment (definitely to the chagrin of Darwin), and it later went into many printings, the best-known version being the second edition of 1845, revised by Darwin himself. It has never been out of print, unlike FitzRoy's volumes, which languished until the re-publication of selected passages in a Folio Society edition in 1977.[10] Darwin's volume had actually been completed in 1837, and even set in type by the end of that year, but had had to wait almost eighteen months for FitzRoy, with his much greater workload on the book alone, and his work on the charts as well, to catch up. And even FitzRoy must have found the life of a newly married man (which had included, as was usual in those days, the prompt arrival of a baby, Emily, on 28 December 1837) some distraction from his work.[11] We get a rare glimpse of that life in a letter Darwin wrote on 1 April 1838:

I went to the Captains yesterday evening to drink tea. – it did one good to hear M[rs] FitzRoy talk about her baby; it was so beautiful & its little voice was such charming music. – The Captain is going on very well, – that is for a man, who has the most consummate skill in looking at everything & every body in a perverted manner. – He is working very hard at his book which I suppose really will be out in June. – I looked over a few pages of Captain Kings Journal: I was absolutely forced against all love of truth to tell the Captain that I supposed it was very good, but in honest reality, no pudding for little school boys, ever was so heavy. – It abounds with

> Natural History of a very trashy nature. – I trust the Captain's own
> volume will be better.

There are a couple of things in the letter – apart from the glimpse into FitzRoy's domestic life – worthy of comment. The remarks about Captain King's journal have sometimes been quoted as implying a criticism by Darwin of FitzRoy's writing, when it is clear from the context that he is commiserating about the amount of work that FitzRoy is having to do to knock King's material into shape. And the reference to FitzRoy's perversity, although tongue in cheek, is probably an echo from a row between the two men that had occurred in November 1837, when FitzRoy read Darwin's draft manuscript of his own volume of the book. He was extremely upset that the younger man had failed in his Preface to give due acknowledgement to the efforts of all the officers on the *Beagle* who had assisted Darwin in so many ways – not least by giving up their own opportunities for recognition as collectors and naturalists. It seems that a great deal of hot coffee was poured before Darwin (who was guilty of an oversight rather than deliberately seeking all the glory for himself) rewrote the offending passage into the form which appeared in print. In telling the story from Darwin's point of view, many accounts give the impression that FitzRoy was more to blame than Darwin in this storm in a teacup; but in the definitive biography, even Darwin's biographer, Janet Browne, accepts that the fault mostly lay with Darwin: 'Darwin did not like admitting he was wrong. He found it much easier to think of FitzRoy as irritable and touchy. He never mentioned the argument again except in his autobiography, and there only to emphasise FitzRoy's uncertain temper.'

FitzRoy was undoubtedly correct, but equally certainly less than tactful in pointing out Darwin's error. As usual with FitzRoy, however, the storm had soon blown over, and he would no doubt have been astonished that Darwin brooded over it so much that he gave it that spin in his autobiography.

Since we have used so much of the *Narrative* as the source material for previous chapters, there is no need to comment further on it here, except that FitzRoy's own volume ends rather oddly, with two supplementary chapters – one 'Remarks on the early migrations of the human race' and the other 'A very few remarks with reference to the deluge'. Mixed up with his own interpretation of Biblical evidence for the differences in skin colour between the different races, the first of these chapters includes a surprisingly modern-looking account from a sailor's perspective of how humankind could have spread around the world from Asia Minor (the main difference from modern ideas, of course, being that FitzRoy perceives these migrations taking place within the past 6,000 years – since 4004 BC – by people created essentially as they are now, while modern scientists see such migrations as taking place over much longer intervals and involving not just *Homo sapiens* but ancestral forms which have evolved into modern human beings). The second of these additional chapters presents a literalist's interpretation of the Biblical story of the Flood, and was clearly written after FitzRoy became aware of some of the geological evidence about the antiquity of the Earth that Darwin intended to include in his own volume on the voyage of the *Beagle*.

The importance of these chapters to us is that they clearly show the pronounced shift in FitzRoy's religious position, from a more or less conventional Christian to a fundamentalist with strongly held belief in the literal truth of the Bible, in the months between the return of the *Beagle* to England and the completion of the *Narrative*. It is, indeed, at the beginning of the chapter on the deluge that FitzRoy tells us how he had in former years 'a disposition to doubt, if not disbelieve, the inspired History written by Moses'.[12] Clearly referring to a conversation with Darwin on the expedition up the Rio Santa Cruz, he says:

> While led away by sceptical ideas, and knowing extremely little of the Bible, one of my remarks to a friend, on crossing vast plains composed of rolled stones bedded in diluvial detritus some hundred feet in depth, was 'this could never have been effected by a forty days' flood,' – an expression plainly indicative of the turn of mind, and ignorance of Scripture.

FitzRoy must have made that remark to Darwin around the end of April or in early May 1834, almost three years into the voyage, and the earliest opportunity he would have had to do much to rectify his 'ignorance of Scripture' would have been two years later, during the quieter moments of the passage home from Australia – when that is, he was not busy writing up his journal.

It is impossible to see any other reason for his 'born again' experience except for the influence of his wife, Mary, who encouraged his Bible reading – he refers to his former lack of 'any acquaintance with the volume', and says that 'for men who, like myself formerly, are willingly ignorant of the Bible, and doubt its divine inspiration, I can have only one feeling – sincere sorrow'. This looks like another clear allusion to Darwin, and whatever the reason for the change, his beliefs would have a profound effect on the later relationship between the two men. But there is no evidence of any profound religious disagreement between them on the voyage itself, and in the short term, by the time the *Narrative* was published in May 1839, their paths had diverged as Darwin settled into his scientific life while FitzRoy showed signs of settling down in England as an aristocratic gentleman and taking up the family tradition of public service.

Even after 1839, as we have seen, FitzRoy was still working on the charts, although he was no longer being paid for this work by the Admiralty. His achievements in command of the *Beagle*, eventually producing eighty-two coastal sheets, eighty plans of harbours and other material including sailing directions for South American waters, had been recognised by the award of the Gold Medal of the Royal Geographical Society in 1837, and in 1838 he was elected as an Elder Brother of Trinity House, the lighthouse authority for England and Wales. The same year, he was also consulted by a select committee of Parliament looking into the problem of what to do about New Zealand, as someone who had actually visited the place and had an interest in missionary work. All of this would have been enough to keep most people busy. But FitzRoy needed more, and with no apparent prospects of another seagoing command, he seems to have responded with alacrity to an opportunity that opened up in the summer of 1841.[13]

By then, Victoria had been on the throne for four years, having become Queen at the age of eighteen on the death of her uncle, William IV. These had been turbulent years politically, with high unemployment and an economic depression leading to the rise of the Chartist movement (a working-class movement for political reform, demanding such controversial things as one man, one vote) and an armed uprising (quickly squashed) in Newport in 1839. Throughout these difficulties (and with the added problem that the image of the Crown had been severely tarnished by the antics of Victoria's Hanoverian predecessors), Victoria had depended on her Prime Minister, the Whig Lord Melbourne. But in 1841 the Whig government fell, and there was a general election. One reason why the Chartists were agitating for reform was that, despite the Reform Act of 1832, many Parliamentary seats were still virtually in the gift of major landowners, with only a few electors, no secret ballot, and possibly dire consequences for tenant farmers who did not vote for the candidate of their landlord's choice. What could have been more natural than that the then Lord Londonderry, an influential member of the Conservative party (as the Tories were beginning to be known), would offer his nephew, Captain Robert FitzRoy, the chance to stand for Parliament in County Durham, in place of a Tory member who was retiring?

The Durham connection had come about in 1819, when Lord Charles Stewart, as he then was (the brother of FitzRoy's mother), a forty-one-year-old widower, had married Frances Anne Vane Tempest, a nineteen-year-old heiress with large estates in the northeast of England that included huge reserves of coal. She brought to the marriage (where she was given away by the Duke of Wellington) the almost unbelievable income of £60,000 per year (close to five million pounds a year in today's money), and at a stroke Lord Charles became one of the richest men in England. He also took the surname Vane, and when he succeeded his half-brother (Castlereagh) as Marquis of Londonderry took to signing himself 'Vane Londonderry'; the descendants of this branch of the family call themselves 'Vane-Tempest-Stewart'. The offer of the seat in Durham really must have seemed like a gift at first, since Durham returned two Members of Parliament (one Whig and one Tory last time around) and only one candidate had been put forward from each party, making the election more of a farce than a formality. But then the kind of trouble which seemed to follow FitzRoy around blew up.

Even within the major political parties, there were different factions (when has there ever been a major political party without factions?), some of whom seemed more interested in fighting each other than their official opponents. Very much against the wishes of Lord Londonderry, a second Conservative candidate, a twenty-six-year-old man called William Sheppard, was adopted as a candidate for the Durham constituency. This was by no means a disaster for FitzRoy. It was always possible that the electors might return both Conservatives and reject the Whig (or Liberal, as they had been known since 1832) candidate; but even if the Liberal did get in, it was highly likely that the candidate backed by Londonderry would come out as the top Tory. The essential thing was that the two Conservative candidates should work together and present a public face of unity. This they did, until (either out of malice or stupidity) a voter revealed that he had instructions from his landlord (Londonderry) to vote for FitzRoy, but not for Sheppard. Although FitzRoy denied that he had any involvement in such shenanigans (assuming such shenanigans were going on), and anyone who knew FitzRoy could not have doubted his word, the young Mr Sheppard hurried off to London to seek advice from an older friend, a Mr Urquhart, who was an anti-Londonderry Conservative and candidate for Sheffield. It was Urquhart who spread stories (almost certainly false) that Londonderry was actually advising the voters of Durham to vote for the Liberal rather than for Sheppard.

Virtually all of the resulting dirty linen was washed very publicly by both FitzRoy and Sheppard, who each published a pamphlet setting out their version of what happened next.[14] Sheppard felt that he had to make a public attack on Londonderry's activities, and that he could not do so while still a candidate for Durham, so he withdrew his candidacy and went off to Sheffield to help Urquhart's campaign. FitzRoy should have kept quiet and got on with his own electioneering; but that was never his way. He saw this as a 'desertion of the conservative cause' and referred to Sheppard's behaviour, in a speech reported in the press, as 'a disgraceful desertion' and 'an event unparalleled in the annals of electioneering'. The only way FitzRoy could really have been harmed by the desertion would have been first for another Conservative candidate to come forward, and then for that candidate to find enough anti-Londonderry support to do better than FitzRoy in the polls. This was never really on the cards, and FitzRoy was duly elected. But meanwhile the feud with Sheppard blew up out of all proportion.

Sheppard took exception to the language used by FitzRoy in his public attack on Sheppard's character, and there followed an angry exchange of letters culminating in Sheppard challenging FitzRoy to a duel. This was still (just) a means gentlemen could use to settle questions of honour; less than a year earlier, Lord Cardigan had wounded an opponent in one of these affairs, and although tried had been formally acquitted of having committed any crime. So FitzRoy had to take the challenge seriously. Seconds were appointed (a Lieutenant Colonel Taylor to act for Sheppard; a Major Chipchase for FitzRoy) and negotiations began about the time and place. But both parties then seem to have realised that things had gone too far, and that they ought to try to find a way out without either of them losing face. Sheppard's second, Colonel Taylor, met FitzRoy in London, and the upshot was a letter from Taylor to FitzRoy which attempted to defuse the situation, but noted that FitzRoy had gone too far in accusing Sheppard of Chartist sympathies. FitzRoy wrote back that, yes, he had been wrong to lump Sheppard in with the Chartists. If only the letter had stopped there – but he went on, in typical fashion, to reiterate his complaints about Sheppard's behaviour. Although the letter ended 'anger does not last with me long', and anyone who had sailed with FitzRoy would have known now to leave well alone, the damage had been done again.

At this point, Taylor decided he had had enough, and left them to it, pleading that he had urgent business to attend to. Sheppard found another second, a Mr Stanley, and together they called on Major Chipchase (a clear breach of the rules of duelling etiquette, since the principals should never be present at a meeting of seconds) and demanded an apology. They were refused. Sheppard decided to try again, in the proper way, and sent Stanley to Chipchase with a formal proposal that apologies should be made on both sides. Chipchase was staying at the same hotel in Durham as FitzRoy, and promptly committed the same breach of etiquette by inviting FitzRoy to join the seconds to discuss the issue.[15] Over a glass of wine, they agreed a form of words, and FitzRoy sent Stanley back with a less than fulsome apology. Sheppard sent him back again to FitzRoy, demanding not only a more strongly worded apology but agreement to publish the correspondence. FitzRoy signed the apology, but refused permission to publish. Everything now escalated yet again, with Sheppard, who seems to have been a remarkably silly young man,[16] accusing FitzRoy of getting Stanley drunk in order to pull the wool over his eyes. By now, Chipchase had also had enough, and withdrew from the affair. Sheppard's tone became increasingly hysterical, denouncing FitzRoy not only as 'a liar and a slanderer' but as 'a coward and a knave'. On the advice of his friends, FitzRoy for once controlled his temper and did nothing. It was a storm in a teacup, and the bottom line was that FitzRoy was an MP and Sheppard was not. But Sheppard still wanted the last word.

On 25 August 1841, FitzRoy was in the Mall, outside the United Services Club, when Sheppard, who had been lying in wait, came up to him brandishing a whip and shouting out, 'Captain FitzRoy! I will not strike you. But consider yourself horsewhipped!' FitzRoy had no such compunctions. After first striking out at Sheppard with his umbrella, he dropped the 'weapon' and set to with his fists, felling his antagonist. The two men never met again, but the affair stirred a flurry of self-justifying letters to the newspapers by the participants, culminating in the two pamphlets we have mentioned. But none of this seems to have provided more than a momentary check on FitzRoy's new career in Parliament, where, as expected (and much to the annoyance of Queen Victoria), his Conservative party now formed the government, under the premiership of Robert Peel.

FitzRoy had no intention of treating Parliament the way his father had, and made a distinctive mark during his brief career as an MP. He was active in the House of Commons, careful to speak only on matters where he had some knowledge (especially nautical matters, but also colonisation, the question of free trade, and the problems of dealing with poverty) and seems to have avoided any major loss of temper. In 1842, he was appointed as a Conservator of the River Mersey, and carried out an inspection of the river (an important trade route for Britain) which amounted to a mini-survey, issuing an eight-page report about the problems caused by the tides and currents in the river, and proposing ways to reduce erosion. He also served on various committees. While all this was going on, Mary FitzRoy gave birth to their second daughter, Fanny, on 27 April 1842. FitzRoy had become a pillar of the Establishment, a respectable and respected MP with a happy marriage, a growing family and bright prospects.

His main contribution as a backbench MP was to bring forward a Private Member's Bill 'to require and regulate the examination of all persons who wished to become masters or chief mates of merchant vessels', which he introduced to the House in a speech made on 28 July 1842, when he was just thirty-seven years old.[17] The Bill proposed to introduce for the first time a system of examining boards, based in the main ports of Britain, to issue certificates that would be a legal requirement for senior ship's officers of all British-registered vessels. He was clearly acting as a stalking horse for the government (a standard role for a backbench MP being groomed for higher things) since the draft Bill went into greater detail than even FitzRoy could have done on his own, and after FitzRoy had spoken the then President of the Board of Trade, William Gladstone, rose to give his commendation to the proposal.[18]

As FitzRoy emphasised in his speech, the object of introducing the Bill at that time (shortly before the summer recess of Parliament) was not in the expectation that it would be passed into law, but to provide a basis for discussion in the House and outside, so that a better Bill, revised in the light of informed comment, might be passed at a later date. It was FitzRoy's Bill that formed the basis of the introduction in 1845 of a system of voluntary certification, followed by the Mercantile Marine Act of 1850 that at last made it compulsory for senior merchant officers to hold such a certificate of competence. Although he was clearly not alone in these endeavours, FitzRoy thereby made a major contribution to preserving human lives at sea, which must have given him the greatest satisfaction.

FitzRoy was next picked out for another job which showed his high standing at the time, although it was exactly the kind of job which his aristocratic background and the broadly based training provided by the Royal Naval College all those years earlier had prepared him for.

Archduke Ferdinand of Austria made an extended visit to Britain in the last months of 1842, touring the country while the ship on which he had travelled was having repairs carried out by the specialists at the Royal Navy Dockyard in Portsmouth. The tour took in Glasgow, Liverpool, Manchester, Chester, Birmingham, Oxford, London, Brighton and Portsmouth, where the Archduke rejoined his ship. The people he visited included the dowager Queen (William IV's widow, who had been so kind to Fuegia Basket), the Prime Minister and the Duke of Wellington (now seventy-three years old). FitzRoy was chosen to act as aide-de-camp for the Archduke throughout these travels, a role which he filled from early October until just before Christmas 1842.

It must have been a very satisfactory Christmas, quite apart from the pleasure of being back home with his family. FitzRoy was established as one of the bright young men of the Conservative administration, and had also made his mark, in a modest way, in diplomatic circles. We shall never know just how far his Parliamentary career might have taken him, but it is certainly not going too far to suggest that he might have achieved Cabinet rank. He would undoubtedly have made a good President of the Board of Trade, for example, and Gladstone's career shows how far that could lead (although we doubt, somehow, that FitzRoy would ever have been a success as Chancellor of the Exchequer). Even without achieving the level of Foreign Secretary or Prime Minister, we can imagine FitzRoy's career leading to a knighthood, perhaps a peerage, possibly to ending his days as First Lord of the Admiralty, which would have been a delightful twist. Even if none of this came to pass, he still (as long as he stayed in England) had a guaranteed income for life from his posts with Trinity House and as Conservator of the Mersey. But none of this was to be. Just when FitzRoy seemed to have found his niche ashore, with his troubles behind him, a settled home life with a growing family and in the foothills of a promising parliamentary career, an offer came along that he could not refuse. Not because it was a wonderful offer in terms of pay, conditions, or prospects, but because it was a job that had to be done, for the benefit of others, and he felt that he could do it. Why give up the possibility of the glorious future we have just outlined, a possibility he must have been aware of, to travel to the other side of the world to carry out a thankless task? For one reason only, the reason that underpins FitzRoy's entire working life. Duty. As the *Good Words* obituary put it, 'to do right, at whatever cost to himself, was the governing principle of his conduct'. That sense of duty was now to lead him into the most difficult period of his life, as Governor of the infant colony of New Zealand – although he set out with high hopes, expressed in a letter to his old commander Captain King, where he said 'my wife and I go willingly, trusting in a superintending Providence – and anxious to raise the New Zealanders. I anticipate no great difficulty with them – but abundant trouble with the whites.' On the second count, at least, his expectations would prove correct.

CHAPTER SEVEN

Difficulties Down Under

FITZROY'S DIFFICULTIES IN NEW Zealand can be understood only in the context of the rather brief history of European involvement with those islands. It was not until 1642 that the land was first seen by Europeans, on board a ship of the Dutch East India Company captained by Abel Janszoon Tasman. It had, indeed, been only about 800 years earlier that the ancestors of the Maoris first arrived on North Island. Tasman thought that this must be an extension of the great southern continent which geographers of the time believed lay in the region, and named it Staten Land, but made no landing, and no attempt to claim the land for Holland. So it was left for Captain James Cook[1] to carry out a proper survey of what had by then been renamed New Zealand. He arrived in 1769 (coincidentally, when FitzRoy's grandfather, the then Duke of Grafton, was Prime Minister), and spent almost six months charting the shores of New

Zealand and establishing that it was actually made up of two large islands and a scattering of smaller ones. He also landed, and formally claimed possession of the islands in the name of George III. This was not strictly in accordance with his orders, which allowed him to take possession of previously undiscovered lands, but although the British government did nothing to formalise this claim and have it recognised internationally, it did establish the British interest in New Zealand.

Cook was in no doubt about the value of the islands. He wrote in his journal that 'the opinion of every body on board [was] that all sorts of European grain fruits Plants &c would thrive here. In short was this Country settled by an Industrious people they would very soon be supply'd not only with the necessarys but many of the luxuries of life.' There was the minor problem of the large population of Maoris, a warlike people who might not take kindly to having their land taken over; but Cook was convinced that they could be won over by 'gentle usage' and by playing one tribe off against another – the old principle of divide and rule.

The only people who took much notice of all this were the members of evangelical and missionary movements, who saw an opportunity to convert a whole population of heathens to Christianity and to bring them the benefits of civilisation. One of the leading lights in setting out plans for such a project was Benjamin Franklin, who wrote in 1771, not long after Cook had returned home, about the possibility of conveying 'the conveniences of life, as fowls, hogs, goats, cattle, corn, iron, etc., to those remote regions', with the intention 'to bring from thence such productions as can be cultivated in this kingdom to the advantage of society'.[2] Nothing came of such plans, not least because Britain and America were soon involved in the American War of Independence. But once the dust had settled on that conflict, with the American continent largely out of bounds, Britain had to look further afield for a dumping ground for its criminals and other undesirables, which led to the formal establishment of the colony of New South Wales in Australia in 1788, and the development of the penal settlement at Botany Bay. To a government prepared to send criminals to the other side of the world to serve their punishment, the distance across the Tasman Sea from New South Wales to New Zealand, a little over a thousand miles, was a mere trifle, and the commission of the Governor of New South Wales extending his authority over 'all the islands adjacent in the Pacific Ocean' was taken to include New Zealand as well. So from 1788, New Zealand became, at least on paper, a British Protectorate, although not yet a British possession.

While Europe was racked by the Napoleonic Wars, New Zealand at last began to feel the impact of European civilisation, as first sealers and then whalers plundered its shores for the benefit of the colonists in New South Wales, and then trade in cargoes such as timber, potatoes, pork and flax became established, with settlers (known as *pakehas* to the Maoris) taking the initiative in building bases there, particularly in the Bay of Islands region, near the northern tip of the North Island of New Zealand. The result was the development of a typical lawless frontier community of the kind epitomised in so many westerns, including brutal repression of the Maoris in some places, and their bloody revenge. Settlers did go through the formality of purchasing land from the Maoris, rather than taking it by force – but, as has almost always been the case in such circumstances, the payments were minimal, and it is not clear that the sellers regarded the concept of ownership of the land in the same way that the settlers did.

New South Wales was too far away, and its resources too limited, for the Governor to exercise any real authority, and the first real civilising influence came with the arrival of the first missionaries, in 1814. Apart from their direct influence for good in the islands, the missionaries would play an important part in the political development of New Zealand because of their strong links with Britain, where their parent missionary societies had great political influence. With Britain still reluctant to take (or accept) full responsibility for New Zealand, the government of the day might well have been glad to see some other power take on the task of bringing law and order to the rough mixture of escaped criminals, deserters from various navies and freebooters on the islands; but they could not let any foreign power (especially a Catholic power) interfere with the lawful activities of the Protestant missionary societies.

Although several abortive proposals were made for colonisation of New Zealand (and one private expedition actually got there, but found such a state of anarchy that they withdrew to Sydney), the overall situation deteriorated into the 1830s, not least since many of the Maoris were now armed with guns, and engaged in fierce fighting among themselves. There were also rumours that the French and/or the Americans were becoming interested in New Zealand. In order to force the hand of the British government, in 1831 (the same year that the *Beagle* set sail on the Darwin voyage) officials of the Church Missionary Society persuaded thirteen friendly Maori chiefs to send a petition to the King (by now, William IV) asking him 'to become our friend and guardian . . . lest strangers should come and take our land'. On the legally nonsensical grounds that these chiefs represented the 'government' of an independent New Zealand state, the British appointed a Resident to take responsibility for, and look after the interests of, the British subjects in New Zealand and establish friendly relations with the Maoris. This was the ineffectual James Busby, who FitzRoy and Darwin would meet on their visit to New Zealand – but, to be fair to Busby, it is hard to see how he could have been anything other than ineffectual, in the circumstances. It would have been better to have bitten the bullet and established a claim of sovereignty immediately; but it has to be borne in mind that the Residency system had worked quite well in India, and that at this time Britain was in a political turmoil which would lead to a real fear of revolution and see troops keeping order on the streets of London. New Zealand was not at the forefront of any British politician's mind.

The appointment of a Resident made little practical difference. The situation at this time can be summed up in the words of a Dr S. M. D. M. Martin, who visited the Cloudy Bay settlement and wrote home:

> If there be a Pandemonium on earth, it must be constituted by the settlement of a number of whaling gangs in the midst of a native population. The Europeans are, as a matter of course, vicious and abandoned; but they have made the natives of Cloudy Bay equally so.

By the end of the 1830s, there were a couple of thousand Europeans (mostly British) living in New Zealand, with a Maori population of at least 100,000, reduced from 200,000 or more twenty years earlier by tribal wars (aided by firearms) and disease (introduced by Europeans). It had become increasingly clear, even to an administration preoccupied by home affairs, that Britain would have to take control of the situation by asserting sovereignty over the islands. In 1838, as we have mentioned, FitzRoy was among those asked by a parliamentary committee for his views on the situation, both because of his visit to New Zealand on the homeward leg of his circumnavigation and because of his missionary interests. 'The Time appears to have come,' he said. 'The Missionaries have opened the Way; they have executed their Office; and now a more sufficient and secure Power ought to step in.' Other people thought so too, and not just in England. By the second half of the 1830s, American ships dominated the New Zealand-based whaling industry, and as early as 1836 a group of the whaling captains and shipmasters had requested their government to provide a presence at the Bay of Islands; the request was granted in October 1838, when an American Consul was appointed, and the flag of the United States was raised officially at the Bay, provoking rumours that the US government had plans to purchase land and start a colony there. Fortunately for the British, however, the Americans (still largely a union of eastern states) were too busy opening up the west of their own continent to contemplate overseas expansion at this time.

More disturbingly, as far as the British should have been concerned (although they seem to have been remarkably complacent at the time), the French had already established a Catholic mission in New Zealand, and in August 1838 a French whaling captain, Jean Langlois, purchased a large area of land from the Maoris in the region of Akaroa, on the South Island of New Zealand. Back in France, he used this as an inducement to persuade a group of investors to form a company to exploit this territory and open up trade with New Zealand. The French government gave the scheme moral support, and something more tangible in the form of an old warship, the *Comte de Paris*, for their first expedition. But they were just pre-empted, when the British government was prodded into action, partly by the actions of yet another independent adventurer, Edward Gibbon Wakefield.

Wakefield was actually something more than an adventurer – but an adventurer he certainly was, having been involved in a famous scandal in the mid 1820s. A senior member of a large and influential family, he had been born in 1796, and after a largely misspent youth eloped in 1816 with a sixteen-year-old heiress, Eliza Pattle. They married in Edinburgh on 27 July that year, and again in London, some time in August. Using the powers of persuasion for which he was to become famous (he was said to have literally hypnotic powers), Wakefield talked his way out of trouble, and secured a generous financial settlement from the marriage. Eliza bore him two children, Susan Priscilla in 1817, and Edward Jerningham in 1820, but died soon after the boy was born. In 1826, Wakefield, still only thirty, seems to have decided that the trick that had worked ten years before was worth trying again, and conned a fifteen-year-old schoolgirl into running away with him to Gretna Green. His plan was to marry her in order to secure the support of her father, a County Sheriff, in entering political life. He had never even met the girl when he managed to fool her into leaving school by sending messages that her mother was seriously ill, then waylaid her in Manchester where he claimed to be a friend of her father, telling her that her father's business had collapsed and that in order to save his remaining property she had to marry Wakefield (as she was under age) so that what was left could be put in his name on her behalf. They married at Gretna Green on 8 March 1826, and Wakefield took her off to France, where the girl's uncles caught up with them and rescued her, without the marriage having been consummated (about the only thing to be said in Wakefield's favour about the whole business). The marriage was annulled by a special Act of Parliament, and Wakefield sentenced to three years in prison for his crimes, along with his brother William, who had aided and abetted him. It was while in prison that he read several books about economics and sociology, and developed his own ideas about colonisation, emerging in 1830 as a seemingly reformed and now respectable character. Doubts about his real motives lingered, however, and prevented Wakefield from fulfilling his wish to enter Parliament. All the evidence suggests that he did not experience a genuine spiritual conversion while in prison, but paid lip service to a package of ideas that allowed him to restore a veneer of respectability to his appearance; as one political opponent later said,[3] the only security against Wakefield was to hate him intensely. So it was natural that he should try to establish the power base denied to him in England in one of the colonies.

From 1836 onward he had been campaigning vigorously to get the authorities to regularise the situation in New Zealand. There was an overt element of idealism in his proposal that instead of England dumping criminals and other riff-raff in the Antipodes, New Zealand should be colonised by the best kind of Englishmen (and women), with the subtext that (rather as actually happened with the British Raj in India) these white gentlefolk would, of course, be supported by a working class of blacks. But at least he recognised that 'the New Zealanders are not savages properly speaking, but a people capable of civilisation', even if he did go on to say that 'A main object will be to do all that can be done for inducing them to embrace the language, customs, religion, and social ties of the superior race.'[4] In 1837, Wakefield formed the New Zealand Association, which later metamorphosed into the New Zealand Colonisation Company, with the aim of putting these ideas into practice, but this was strongly opposed by the Church Missionary Society, then at the height of its power and influence, which wanted a free hand to bring Christianity to the Maoris. With pressure from just about all sides, and reports of the failure of the Residency system to bring order to the chaos in New Zealand, the government at last took action, in December 1838.

Their first action, still regarding New Zealand as an independent sovereign Maori state, was to appoint a British Consul for New Zealand, in the form of a Navy Captain, William Hobson, who had experience of the Pacific, but was at that time on leave in England. After taking advice from the Law Officers of the Crown, in June 1839 the British government formally extended the authority of the Governor of New South Wales to include 'any territory which is or may be acquired in sovereignty by Her Majesty, Her Heirs or successors, within that group of islands in the Pacific Ocean, commonly called New Zealand'. So when Hobson sailed from England at the end of August 1839, he carried with him the documents extending the authority of the Governor of New South Wales, his own appointment as Consul, and another commission appointing him as Lieutenant Governor of whatever territory he could persuade the Maoris to cede formally to the British Crown. A further document set out new rules for the purchase of land from the Maoris – no legal title would be recognised unless the land was first purchased by the Crown before being passed on to the settlers. Although it was recognised that the North Island could be ceded to the British only by a legal agreement with the chiefs, there was some confusion about the status of South Island, which the British authorities in London wrongly thought to be inhabited by rather less civilised tribes – in fact, if anything the South Islanders were more civilised. Because of this, Hobson had the authority, if he felt it appropriate, to claim sovereignty over South Island by right of discovery.

The new Consul and prospective Lieutenant Governor arrived in New Zealand (having stopped off in Australia along the way) at the beginning of 1840. The situation there was complicated by the fact that the advance guard of Wakefield's group, including his son (then nineteen) and his brother, had set out from England on 12 May 1839, on board the *Tory*. They left without government approval, and had arrived in New Zealand in August 1839, shortly before Hobson had left England (by a quirk of fate, the ship was commanded by Edmund Chaffers, who had been FitzRoy's sailing master on the *Beagle* during the second surveying voyage to South America). But the trouble that would cause still lay ahead as Hobson set about his task of claiming New Zealand for the British Crown (in the person of Queen Victoria, who had come to the throne in 1837).

The proceedings were something of a legalistic farce, still being based on the assumption that the chiefs of the Maori tribes represented the islands' sovereign government. On 6 February 1840, just forty-five chiefs signed the Treaty of Waitangi, recognising British sovereignty but granting the Maoris 'all the Rights and Privileges of British subjects'.[5] Then, copies of the Treaty were carried around the North Island to be signed by other chiefs, with British sovereignty over the whole of North Island being proclaimed on 21 May. Hearing reports that the French expedition was in New Zealand waters, Hobson then used the discretion allowed for in his orders to claim the South Island for Britain, by right of Captain Cook's discovery. As an insurance, several chiefs on the South Island were also induced to sign the treaty, and sovereignty over South Island was proclaimed on 17 June 1840. The whole of New Zealand was at last British, as far as international law was concerned[6] – much to the disappointment of the French settlers on board the *Comte de Paris*, which arrived at Akaroa in August 1840. If they had arrived a year earlier, South Island, at least, might have become a French colony; instead, the small group of settlers made their new homes on British soil. After Hobson's dispatches reached London, the appropriate proclamations were formally published there on 2 October 1840, the final act in the establishment of British sovereignty over New Zealand.

Even before the various chiefs had added their signatures, or marks, to the Treaty of Waitingi, however, Hobson had been taken ill, struck by a paralysis 'occasioned by harassing duties and by long exposure to wet'. Sir George Gipps, the Governor of New South Wales and therefore now in overall charge of the New Zealand colony, was sufficiently concerned by this to send a detachment of eighty troops, under Major Thomas Bunbury, to take over in New Zealand if Hobson was incapable of governing (Hobson had arrived with a token force of just four constables to back up his authority), but by the time they arrived Hobson's health had improved.[7] Even with improved health and this force to back him up, though, he found it increasingly difficult to exert his authority. The United States never did recognise his position officially, the French only grudgingly accepted the fait accompli, the Maoris never had been united in agreeing to the British takeover, and the arrival of Wakefield's expedition split the British colonists between the already established missionaries and their associates, who tended to support Hobson as the least of several evils, and the newcomers, who chafed against the constraints imposed on them, particularly in the matter of purchasing land. If real trouble arose, the force at Hobson's disposal would be insufficient to police even the colonists, let alone the tens of thousands of warlike Maoris.

The main bone of contention was, of course, land. Following Wakefield's lead, more and more colonists arrived, mostly settling in the region either side of the Cook Strait, between the two main islands of New Zealand, laying the foundations of (among others) the communities of Wellington and Nelson. Hobson established his administration further north, in Auckland, near to the original missionary settlements, which was seen by the newcomers as identifying him with that faction, and by implication with what the new settlers saw as an over-protective attitude to the Maoris. The newcomers wanted to buy land, and the Maoris wanted to sell it; but under the new regulations only the Governor, on behalf of the Crown, could purchase land, and he had extremely limited funds at his disposal. What land he did buy was purchased at a price which the Maoris thought to be too low, and they also (not unreasonably) objected when the Crown sold land on to settlers at a profit. The situation was exacerbated because the wording of the Treaty of Waitingi introduced to the Maoris the idea that they owned all of the land, including unoccupied territory; previously, they had regarded unoccupied land as free for anyone to move into, whether it be a Maori tribe or white settlers. Arguably, Hobson's biggest mistake was not simply to claim all the unoccupied land of New Zealand for the Crown, and his critics were not slow to point this out. As if all that were not enough, Hobson's authority was initially reduced by his position as a mere Lieutenant Governor, subordinate to Sir George Gipps (and there were even grumblings from the New Zealand colonists about the lack of status implied by being legally tied to the convict settlement in New South Wales). This indignity, at least, was removed on 16 November 1840, when New Zealand was granted the status of a separate colony in its own right. Hobson became the first full Governor and Commander-in-Chief of the Colony of New Zealand, with powers that were essentially unlimited, especially given the remoteness of the colony from London and the slowness of communications by sea.

This power, though, was in the hands of a man who was still unwell, even though not as ill as he had been, and who lacked any armed force with which to back up his paper authority. The actual day-to-day running of the colony increasingly fell to subordinates, many of whom were self-serving rather than public-spirited, and not a few of whom were undoubtedly corrupt. Within a year, the colony was technically bankrupt; within two years, the administration was widely regarded as incompetent and oppressive, with protest meetings being held not just at the Company's power base around Wellington and Nelson, but even in Auckland, at the seat of government. The seriousness of the deteriorating situation became clear back in London, where on 18 February 1843 the Parliamentary Under-Secretary to Lord Stanley, the Secretary of State for the Colonies, wrote to inform Stanley that 'I am afraid that it will be necessary for you before long to resort to . . . a change of Governors.'

But the decision had already been taken out of Stanley's hands; on 10 September 1842, Hobson had died of a stroke, leaving the bankrupt colony temporarily in the hands of his number two, Lieutenant Willoughby Shortland, Bunbury having returned to his normal duties when Hobson made his initial recovery from the paralysis that afflicted him in 1840. The underlying reasons for the difficulties the colony now faced lay in the confused policy of the British government concerning New Zealand, and their reluctance to either fund the colony properly or police it properly.[8] But a dead Governor makes a good scapegoat, and Hobson's incompetence (partly, though not entirely, a result of his ill health) helped, temporarily, to obscure these deficiencies. As McLintock puts it, 'it was Hobson's tragedy – no less New Zealand's – that his weakness as a man obscured the fundamental weakness of Colonial Office policy', which helped to ensure that no immediate action was taken to address the underlying problems, and ensured that the appointment of a new Governor was seen as the solution to all the problems. But before that new Governor, Robert FitzRoy, could be appointed, there was still time for the situation to deteriorate further under Shortland.

Unfortunately, Shortland was the most unpopular member of Hobson's administration, and as one of the men most responsible for the unsatisfactory state of affairs he was never going to be seen by either group of colonists as the right person to sort out the mess. His interim administration could be only a weak stopgap awaiting the arrival of a new Governor. The financial situation, in particular, continued to deteriorate, as Shortland seemed incapable of coming to grips with what was admittedly a complex and almost incomprehensible system of colonial book-keeping. Following a lead set by Hobson, he kept the colony afloat financially by issuing bills (essentially a form of paper money), without the authority of the government in London – indeed, going against their specific orders; in spite of this the bills were eventually honoured by the Treasury.

In this power vacuum, some impatient settlers decided to take matters into their own hands. A little to the northeast of Nelson, in the Wairau valley, there was a rich region of potential farmland which the Company claimed (and genuinely believed) it had purchased from the Maoris. The local Maoris, headed by two chiefs, Rauparaha and Rangihaeata, claimed that they owned the land, and that no sale had been agreed. A Land Claims Commission set up by Hobson was looking into the dispute, along with many other similar disputes, but taking far too long to reach a decision in the eyes of the settlers. A retired naval Captain called Arthur Wakefield (one of the brothers of Edward Gibbon Wakefield) sent his own surveyors into the valley in preparation for parcelling up the land for individual families, and there they were harassed by the Maoris, who disrupted their work and burnt down one of the huts they were living in. Wakefield responded by 'persuading' a local police magistrate to issue a warrant for the arrest of the two chiefs, and foolishly set out with a raggle-taggle crowd of armed settlers in a posse to enforce the warrant. Since the Maoris also had firearms, were better disciplined and had far more experience of warfare than the settlers, the result was inevitable. On 17 June 1843, after a minor exchange of fire in which three of the settlers were killed and most fled, nineteen survivors asked for a truce under a white flag. Interpreting the white flag as a sign of surrender, the Maoris lined up their prisoners and did what they usually did to prisoners – killed them all, including Arthur Wakefield.

It isn't always appreciated that these events caused as much of a shock to the Maoris as to the settlers – as FitzRoy later put it:

A shock which vibrated through the length and breadth of the land. That the settler should try to take land by force of arms was a startling idea, and it at once revived every former suspicion. Until then the settlers had been supposed to be men of peace, and trade; and the missionaries had invariably done their utmost to prevent warfare, but a new view was opened by the collision at Wairau.[9]

It would have been almost as easy for the Maoris to kill every white person in New Zealand at that time, but whether out of political acumen or because he was simply too frightened to do anything, Shortland refused to bow to the demands of the Company and its settlers for action, which he saw 'must cause a vast and useless sacrifice of human life'. The immediate effect was that the more hostile Maoris were emboldened further, the Company was put on the defensive politically, the supply of migrants dried up, and the older-established missionaries were able to take the moral high ground in their relationship with the politicians. It was at this point that FitzRoy came on the scene. It would have taken a political genius (perhaps someone like his uncle, Castlereagh) to resolve the situation to everyone's satisfaction, and for all his virtues FitzRoy was not a political genius. But he arrived with a fund of goodwill, and he made a realistic attempt to make the colony successful. He failed because he had been given an impossible task; but the nature of his failure was typical of his character. He emerged, in the words of modern New Zealand historian Paul Moon, as 'possibly the most maligned, and certainly the most misunderstood, of all New Zealand's colonial leaders'.

When the news of Hobson's death reached England, the Colonial Office had had little hesitation in offering FitzRoy the post of Governor of New Zealand.[10] There was no need for long deliberations over the vacancy, because the possibility of needing a successor to Hobson had been discussed back in 1840, when the news of Hobson's paralysis had reached them. Even at that time, FitzRoy was mentioned as an ideal candidate for the job, with the influential Church Missionary Society putting his name forward. FitzRoy, of course, was known to them as a devout Christian who not only shared their views on uplifting, or raising, savages by introducing them to the benefits of civilisation and Christianity, but had tried to put these ideas into practice with the Fuegians. In July 1840, Dandeson Coates, the Secretary of the Church Missionary Society, had written to James Stephen, the Permanent Under-Secretary at the Colonial Office,[11] to present the case for FitzRoy:

My solicitude is great that [the next Governor] should be an individual whose primary care should be the welfare in the largest sense of the lives of the Native race. Have you any such person, duly qualified in other respects, in view? If not, what do you think of Captain FitzRoy? I have reason to believe that he is deeply interested in New Zealand ... and would not hesitate to accept the position if it were offered to him. He certainly seems to me to combine many valuable qualifications for an office which is certainly one of much difficulty and delicacy.

And since then FitzRoy's credentials for such a post had been enhanced by his election as an MP (in spite of the unfortunate circumstances associated with that election) and his work in the House of Commons and as aide-de-camp to Archduke Frederick. When the offer did come his way, in the spring of 1843, FitzRoy did indeed have no hesitation in accepting it, even though this meant resigning his seat in the House and giving up his life in England, probably, as he expected at the time, for ever. Fully briefed by the Colonial Office – although not with news of the Wairau massacre, which did not reach England until after FitzRoy had sailed – he can have had no illusions about the mess he was heading into, and it is confirmation of FitzRoy's self-confidence (and perhaps his belief that God was pointing the way for him) that he should have taken such a bold step. It is clear that he did so not out of personal ambition, but because he felt that he was indeed the right man to look after 'the welfare in the largest sense of the lives of the Native race'.

But it took time to sever those ties and make preparations for not just FitzRoy but his whole family (including his wife, Mary, their three small children, and her retired father, Major General O'Brien) to make preparations for the move, and it was not until 8 July 1843, just three days after FitzRoy's thirty-eighth birthday, that they set sail, from Torbay, as passengers on the merchantman *Bangalore*. Their leisurely voyage included a call at Bahia, where FitzRoy delighted in showing his wife some of the sights he had seen when in command of the *Beagle*, before proceeding around the Cape of Good Hope and via Australia to New Zealand. The ship reached New Zealand on Saturday, 23 December, and it was deemed inappropriate for the new Governor to officially take up his post on a Sunday or on Christmas Day. So although the family had attended church on both the previous days, FitzRoy then returned to the ship, and it was on 26 December that he formally set foot in the colony, arriving with a flourish that hardly seemed appropriate to the circumstances, and was mildly ridiculed at the time:

> A gentleman connected with the native department carried a pole surmounted by a crown of flax, from which waved the New Zealand flag; and Captain FitzRoy, excited by the occasion, cried aloud when stepping on shore, 'I have come among you to do all the good I can.' The crowd of fifty persons replied to this noble sentiment with a cheer, and the commanding officer of the company of soldiers in attendance shouted, 'Quick march'; immediately the two drummer boys and the fifer of the guard of honour struck up 'The king of the Cannibal islands' to which appropriate air His Excellency marched to Government House.[12]

In view of the general recognition of FitzRoy's term of office as a failure, and the opinion often expressed in historical accounts that his appointment was a mistake, it's worth noting that the appointment was warmly received both in New Zealand and by, for want of a better term, propagandists for New Zealand back in London. FitzRoy's aristocratic background and powerful connections, his achievements as a navigator and surveyor, and his position as a rising young man in Parliament, seemed to the *New Zealand Journal*, based in London, to indicate that the government was at last taking New Zealand seriously, and appointing a heavyweight to look after its affairs. In New Zealand itself the leading newspaper of the day, the *Nelson Examiner*, was no less fulsome in its delight that the rule of 'ignorance and stupidity' was coming to an end, and expressed pleasure that a man had been appointed who possessed the confidence of 'the entire European population' of the islands. As McLintock points out, it is particularly surprising that the southern colonists around the Cook Strait should have felt this way, since FitzRoy's missionary interests and connections with the Church Missionary Society were well known; but it shows, perhaps, how much the colonists wanted somebody to wave a magic wand and make everything better, as well as their genuine respect for FitzRoy's past achievements.

Of course, there was no magic wand. FitzRoy was faced with three problems – lack of finance, the split between the missionary colonists in the north and the Company colonists in the south, and growing Maori unrest – which simply could not be resolved with the resources at his disposal. His one great asset was an ability to improvise, think on his feet and make decisions without waiting for advice or orders from superiors – all excellent qualities in a naval officer, especially one with an independent command like the *Beagle*,[13] but which would ultimately give London the excuse to recall him. Unfortunately, one of FitzRoy's other characteristics, an inability to suffer fools gladly, got him off on the wrong foot with the one man you might have expected him to turn to for advice about the current situation, the Acting Governor Lieutenant Shortland (whose substantive title was Colonial Secretary). Thomson describes what happened at a levée (a formal party held to introduce the right kind of people to the new Governor) just after the events of 26 December:

> Next day a curious scene occurred at the levee. The colonial office had given Captain Fitzroy files of a New Zealand newspaper, famous for abusing Acting-Governor Shortland, to read during the voyage; and when the editor of that paper was presented at Government House, the Governor informed him that he highly approved of the principles of the *Southern Cross*. This speech, equivalent to announcing in the Government Gazette that the colonial secretary was an arrogant fool, caused Mr. Shortland to resign his office; and Dr. A. Sinclair, a surgeon of the royal navy, who had accompanied Captain Fitzroy to explore the natural history of the country, was appointed colonial secretary in his stead.

Officially, Shortland was dismissed by FitzRoy, and this story may have got exaggerated in the telling; but whether he jumped or was pushed the result was the same. Shortland left in the same ship (indeed, in the same cabin) that the FitzRoys had arrived in, ensuring a complete break between FitzRoy and the previous regime – probably as he intended, but possibly unwisely.

It didn't take long for FitzRoy to antagonise in similar fashion the one group of people who might have been particularly pleased to see Shortland replaced by a new broom – the southern settlers. One of the most outspoken critics of Shortland had been Edward Jerningham Wakefield (usually known as Jerningham to distinguish him from his father Edward Gibbon Wakefield), the nephew of Arthur Wakefield. It was clear that something had to be done about the relationship between the Company settlers and the Maoris in the wake of the Wairau massacre, and as soon as he had settled his family in the north, FitzRoy travelled down to Nelson. Following the death of his uncle, Jerningham Wakefield had written some inflammatory newspaper articles, and had been active in promoting the idea of revenge against the Maoris. One of his choice phrases was that the Maoris ought to be 'crushed like a wasp in the iron gauntlet of armed civilisation'; perhaps not a sensible proposal (whatever its moral repugnance) to put forward in a heavily outnumbered community protected by a few dozen soldiers and confronted by opponents who also had firearms. He soon found that this was not the way to get on with a Governor fired with missionary zeal and a desire to see the savages uplifted into the Christian community. But he might have realised that.

Jerningham's own account,[14] which is extremely biased but is an accurate reflection of the hostility with which FitzRoy's appointment was greeted by the Company settlers, tells us that when the settlers were discussing who might succeed Hobson:

> Captain Fitzroy's name was sometimes mentioned. But that officer was known to be so thoroughly prejudiced in favour of the narrow philanthropy of the pure missionary system, unmingled with the concurrent benefits of civilization, that such an appointment was looked upon as probably subversive of the last hope for the natives. I remember one morning hearing several of the best and bravest settlers, collected in Colonel Wakefield's house, agree, 'that when they heard Fitzroy was Governor, it would be time to pack up their things and go.'

It is against that background that we should read Jerningham's account of the levée at which FitzRoy was welcomed in Nelson, at the end of January 1844, when Jerningham was in his twenty-fourth year. He tells us how after the usual congratulatory address on the safe arrival of the new Governor and his family in New Zealand, as FitzRoy began to speak 'a general stillness, a sort of chill or damp seemed to creep over the noisy bustle of the crowd as his opinions were gradually made known'. After emphasising that 'all parties might rely on receiving justice, and nothing but justice' at his hands, FitzRoy then:

> deprecated, in the strongest terms, the feelings displayed by the settlers at Wellington against the native population, of which he judged by what appeared in their newspapers. He stated that he considered the opposition to the natives to have emanated from young, indiscreet men; but he trusted that as they had years before them, they would yet learn experience ... Having so lately left England, he could not be ignorant of the intention of people there; none would emigrate to New Zealand unless they believed there was a good understanding between the settlers and the natives, and unless the settlers did all in their power to conciliate the natives, to forgive them, and to make allowances for them because they were natives, even if they were in the wrong.

It is easy to imagine how those sentiments were received; a few minutes later, after the speeches had finished, FitzRoy picked Jerningham out of the crowd and in front of the interested bystanders and 'with a frown on his face, and the tone of the commander of a frigate reprimanding his youngest midshipman', said:

> When you are twenty years older, you will have a great deal more prudence and discretion. Your conduct has been most indiscreet. In the observations which I made to this assembly just now, I referred almost entirely to you. I strongly disapprove and very much regret everything that you have written or done regarding the missionaries and the natives in New Zealand. I repeat that your conduct has been most indiscreet.

Following a private meeting at which FitzRoy berated Jerningham (according to Jerningham) in his best quarter-deck manner, the young man decided that he 'could stay no more in the country with comfort under this Government; for so long as Captain Fitzroy ruled, I must always appear to a certain degree as a disgraced member of the society'. He did indeed pack up his things and go, and had left New Zealand before FitzRoy had returned to his seat of government in the north (although he did return, and played a part in the affairs of the colony, after FitzRoy had been recalled to London). Unfortunately, FitzRoy's other problems could not be got rid of so easily.

Before he could return to the north, FitzRoy had to settle the matter of the Wairau massacre and the land dispute that had triggered it. He travelled with his entourage out to one of the Maoris' fortified villages to hear the evidence and pass judgement in the case. He had already discussed the settlers' version of events with the Nelson magistrates who had signed the original warrants for the arrest of Rauparaha and Rangihaeata after the hut-burning incident; now after listening to the Maori side of the story, including their claim that it was their usual custom to kill chiefs captured in battle, he considered the matter for about half an hour, discussing the finer points of the translation with his interpreter. Then he spoke:

> Listen, O ye chiefs and elder men here assembled, to my words. I have now heard the Maori statement and the *paheka* statement of the Wairau affair; and I have made my decision. In the first place the white men were in the wrong. They had no right to survey the land which you had not sold [until the Land Commission had settled the case]; they had no right to build the houses they did on that land. As they were, then, first in the wrong, *I will not avenge their deaths* . . . But although I will not avenge the deaths of the *pahekas* who were killed at the Wairau, I have to tell you that you committed a horrible crime in murdering men who had surrendered themselves in *reliance on your honour as chiefs*. White men never kill their prisoners. For the future let us live peaceably and amicably – the *paheka* with the native, and the Maori with the *paheka*; and let there be no more bloodshed.[15]

There is no doubt that FitzRoy acted out of the highest possible moral motives, as well as seeing the impracticality of armed conflict with the Maoris with the forces at his disposal.[16] But he ended up in an impossible situation. The settlers regarded him as weak and cowardly, while the chiefs were amazed that nothing had been done to punish them or take revenge. Probably, what FitzRoy should have done was to claim the disputed land in compensation for the blood of the settlers, in line with the traditional Maori concept of *utu*. As it was, the Maoris saw a weak opponent who would not fight to defend his rights. But what could he have fought with? As FitzRoy pointed out in his *Remarks on New Zealand*, published after his return to England, if he had tried to arrest the two chiefs and bring them to trial, the ringleaders

> would have retreated into their fastnesses, where no regular troops could have followed: thousands would have joined them: hostilities against the settlers would have been commenced, and their ruin must have followed: – ruin under the most horrible circumstances of heathen warfare.[17]

A little earlier in his *Remarks*, FitzRoy comments on the unrealistic attitude of the settlers in the south:

> They would not believe that the natives could ever become formidable opponents, or that it would be useless to cultivate the soil if only under the protection of troops. They would not believe that no one could work in the interior while continually exposed to the rifle of the native; neither would they believe that no produce of the land could pay for cultivation at the point of the bayonet . . . No one appeared disposed to give the natives credit for courage, or skill in warfare – no one seemed to doubt that they would fly before a very small detachment of artillery. The prevailing feeling appeared to be anxiety for a collision.

A collision, that is, between tens of thousands of Maori warriors, many armed with muskets or rifles, and the 134 soldiers available in New

Zealand to defend seven settlements scattered around the two main islands.

Stanley and his officials, studying reports of the massacre at the Colonial Office, reached much the same conclusion as FitzRoy, at much the same time;[18] FitzRoy had no choice and, as one contemporary later put it, he 'had made a virtue of necessity and affected to concede voluntarily an impunity which he did not dare to refuse'. But perhaps, as we have suggested, he could have done it more subtly.

Having satisfied nobody, the Governor set off back to the north, to come to grips with the colony's desperate financial situation. While he did so, as the Maoris in the north learned what had happened (or not happened) as a result of the Wairau massacre, the hothead element among them (their equivalent of Jerningham Wakefield and his ilk) started to foment a feeling that it was time for the northern tribes to get in on the action as well. The coming of the settlers had long since begun to break down the traditional tribal authority which might otherwise have held them in check, and the replacement for that authority was also crumbling. By this time, the novelty value of Christianity had begun to wear off,[19] and the teaching of the missionaries about the Christian way of life had been severely undermined by the decidedly unChristian way of life of the whalers and their like – although there remained a core of Christian Maoris in the north, providing an example of the kind of success that FitzRoy craved, and enabling Mary FitzRoy to write back to his sister Fanny that 'the natives are certainly a most intelligent, interesting race – many very well dressed in European clothing have been with us at different meals and behaved *perfectly* . . . They appear to understand every measure of gov^t thoroughly.'[20]

If FitzRoy had been free to concentrate on the internal politics of New Zealand, he might have been able to resolve these issues without conflict. But the financial crisis was so pressing that it now took up much of his time, ultimately to little avail. When he was able to make sense of the records (which had not been properly kept) he found that at the beginning of 1844 the colony was in debt to the tune of £24,000, while it had an estimated income of £20,000 a year and heavy commitments, including salaries for a top-heavy administrative bureaucracy. The policy of the government in London was that the colony should stand on its own feet, without financial aid from the mother country; the reality FitzRoy found on the ground was that 'all salaries and ordinary current payments were several months in arrear: there was no prospect of the revenue amounting even to two-thirds of the estimated indispensable expenditure'.[21] The situation was exacerbated because no new immigrants were coming into the country in the wake of the Wairau massacre, so there was no new money, while a collapse in trade (related to Maori unrest) reduced the government's income from customs duties. In addition, land sales had come to a halt under the cumbersome system of the Crown monopoly, which was widely regarded as unfair.

In the best naval tradition, FitzRoy used his initiative, without asking permission from superiors who were so far away that it could take six or eight months for dispatches from London to reach him, and just as long for his reports to reach London – just as, not so long ago, he had not waited for permission before purchasing a schooner to work with the *Beagle*. Directly disobeying his instructions, like Hobson and Shortland before him he issued promissory notes, which were later authorised as legal tender. The snag was that these notes were not backed by any reserves (there was no Bank of New Zealand, with vaults full of gold), and would clearly be redeemed only if and when the British bailed the colony out. So they rapidly depreciated in value, while at the same time everybody hoarded their 'real' money against the inevitable rainy day. But at least there was money circulating (as much as a nominal £26,000 worth of paper currency at one stage), so wages could be paid and business transacted.

In order to encourage trade, in May 1844 FitzRoy proposed abolishing all taxes on commerce, introducing a free trade system throughout the islands.[22] This really was a case where FitzRoy acted impulsively, without thinking things through. Free trade was fine in principle, but the government had to have revenues, and the taxes he proposed abolishing were to be replaced by taxes on houses and land. Again, fine if they could be collected, but in practice impossible to collect if the citizenry refused to pay, which they did. Faced with fierce opposition, FitzRoy withdrew the plan, which meant going back to customs duties as the government's source of income. In spite of howls of protest when these duties were increased, the reforms might have worked if there had been time to collect the revenue and distribute it in wages and so on – in other words, if the British had even lent the colony (let alone given) something like £10,000 to tide them over. But before the money could flow through the system, the economic situation deteriorated further, now linked to increasing unrest among the Maoris in the north, who had got used to the benefits of trading with the settlers and were experiencing an economic recession for the first time. This local recession was exacerbated by an economic depression in New South Wales, by far the biggest market for New Zealand. In this fertile ground opponents of the FitzRoy regime encouraged a growing belief among the Maoris that the collapse in trade and in the standard of living they had begun to be accustomed to was caused by the increased customs duties. In desperation, in spite of all the warnings FitzRoy did go over to a system of free trade, linked with a tax on property and income which proved, in the words of the *New Zealand Spectator*,[23] 'a melancholy farce at which we know not to laugh or cry'. The old system was restored in April 1845, but by then the writing was already on the wall for FitzRoy.

While all this was going on, FitzRoy also had to deal with the land question. There was one particularly trying case, where the restlessness of the natives threatened violence. Around Taranaki, near New Plymouth, the Company had purchased some 60,000 acres of land from the Maoris – but, according to the natives who were now getting restless, the wrong Maoris. What had happened was that at the time the Company came on the scene, this region had been largely swept clean of its previous occupants by tribal warfare (warfare made worse by the availability of guns purchased from the whites). The victorious tribes in this conflict had sold the land to the company, but the original inhabitants had now come back, and were trying to evict the whites in the same way that they had been evicted by the other tribes. The Land Commissioner, William Spain, had ruled that the Company had legal title to the land, whoever they had bought it from. FitzRoy overturned this decision in August 1844, and gave it back to the Maoris. The care with which FitzRoy listened to the Maori arguments showed that he empathised with the importance to them of the form of such issues, not just the ultimate outcome, and represented his attempt to combine their tradition with British law. He tells us that he once witnessed a discussion between two families, lasting nearly a day, over who had the right to a ship which had been bought from an Englishman in exchange for land:

> After tracing back their respective descents through eight generations, eagerly contesting every point, both parties agreed that the actual sellers of the land had not the right, and that the vessel ought to belong to the others, who also were willing to sell [the land], on the same terms. On this the chief of the real owners waived the right of himself and his family, saying that they did not really want the vessel, but they wished their right to be known and acknowledged.

This closely matched the aristocratic code of honour which the FitzRoy family had lived by since the time of Charles II. FitzRoy understood the importance to the Maoris of this kind of acknowledgement of their rights. As with the Wairau incident, on this occasion his decision was the only pragmatic solution to the problem; but it also fitted FitzRoy's moral sense of duty, and should have convinced the Maoris that whatever their other causes of complaint, in FitzRoy they had a champion of native rights. But, of course, it further antagonised the southern settlers.

It's worth pointing out that FitzRoy's high moral stand was not in conflict with his instructions from London. Far from it. He had seemed the right man for the job in no small measure because he held those principles, and was endorsed by the Church Missionary Society. The impression FitzRoy had from the outset that he had the support of the Colonial Office in putting these principles into practice was reinforced by communications he received from Stanley – one dispatch, dated 13 August 1844, spells out that:

> What you and I have to do is to administer the affairs of the colony in reference to a state of things which we find, but did not create,

and to feelings and expectations founded, not upon what might have been a right theory of colonisation, but upon declarations and concessions made in the name of the Sovereign of England [a reference to the Treaty of Waitangi, which gave the Maoris 'all the Rights and Privileges of British subjects']. [The Maori people should be governed] by a conciliatory course, to bring more close connexion with, and more complete subjection to British authority.

How could FitzRoy fail to see this as an endorsement of his own principles? Yet he was still the pragmatic naval officer, and knew that moral persuasion alone would never make honest British citizens out of the more warlike Maori tribes. In his *Remarks* (admittedly, written with the benefit of hindsight), he spells out the reason for this need for pragmatism:

Repeated denials given to re-iterated applications of successive governors of New Zealand for more effective support of their position, obliged them to have recourse to a system of forbearance and conciliation, which – in the nature of things – could not long continue, and which encouraged encroachments, as well as injurious trials of strength, on the part of both races. In the colony an extreme of forbearance – arising out of utter inability to carry out the law efficiently, rather than from real leniency, bordered on inhumanity towards the settlers.

And as an example of those 're-iterated applications' for support, on 19 October 1844, in a dispatch to Stanley he wrote that: 'There are now very strong reasons for the presence of a regiment of the line, and at least two ships of war, in New Zealand. The speedy appearance of such a force may save years of misfortune, misery and bloodshed.' The reasons why he wrote in this way at that time will soon become apparent; for now, it is enough to say that history proved him right. He was also aware that it is also possible to tame a cat by feeding it cream. On 16 September 1844, he had pointed out to Stanley that: 'If I had the means of paying small salaries . . . to the principal chiefs . . . it would be a great hold on their allegiance, and ensure their helping to keep the peace.' But like all his requests for financial assistance, this was ignored.[24]

FitzRoy continued to be forced to try to patch things up piecemeal, tackling problems as and when they occurred, instead of being able to take overall charge of the situation on the basis of having a well-funded administration backed up by a realistically large armed force with which to impose law and order on settlers and natives alike. In the light of these difficulties, FitzRoy's response, according to contemporary sources quoted by McLintock, was one of unswerving devotion to duty and utter unselfishness. He 'made no effort to spare himself in the conduct of public business; at any time, it was said, he would cheerfully sacrifice his private interests and even his reputation at the Colonial Office for what he believed to be the good of the Colony.' In a letter to Robert's sister Fanny, dated 19 December 1844, Mary ends by saying that 'Robert has just come in to bed,' and that he asks her to say that although Fanny knows only too well how hard he has worked all his life, 'it was *nothing* to what he has to do now, & that he has not leisure now to write any private letter – but that he is quite well. His patient cheerfulness often surprises me.'

The mistakes he made were largely those forced on him by the constraints he worked under, and his failure to meet all the requirements of the Colonial Office (notably in neglecting to send detailed dispatches as often as they would have liked) resulted as much as anything from the way his time and energy were swallowed up in finding short-term solutions to one pressing problem after another. As FitzRoy wrote in a note to Stanley on 19 May 1845: 'I have no time to write private letters – even to your Lordship: my whole time and thoughts from five in the morning till the morning again [that is, after midnight], being constantly given – as it ought to be in these critical times – to publick duties.'

In one move which almost came off, FitzRoy tried to solve the problem of his empty exchequer and the blockage on land sales in one cunning step. In March 1844, he abolished the Crown's monopoly on land purchases, and allowed sales to proceed direct from the Maoris (who were eager to sell) to the settlers (who were eager to buy), subject to a tax for the government of ten shillings per acre, and to permission from FitzRoy's administration. As he told the chiefs:

There is no longer any objection to your selling such portions to Europeans, provided that my permission is previously asked, in order that I may inquire into the nature of the case, and ascertain from the protectors whether you can really spare it, without injury to yourselves now, or being likely to cause difficulties hereafter.[25]

We get one intriguing sidelight on the situation at this time from Sir Everard Home, writing to the Hydrographer from his ship HMS *North Star* in Hobart, Tasmania, on 3 May 1844. After reporting several matters to London, in a PS he mentions hearing from FitzRoy who 'I believe gets on very well and writes to me that he wants nothing but money. He does not want force, but he has no money and cannot draw for any. I should give it up.'

Home's laconic comment reveals an astute grasp of which way the wind was blowing. The ploy of abolishing the Crown monopoly cleared the initial logjam, with 600 acres of land being sold to the most eager settlers, and briefly raised hopes that the colony's problems had been solved. But the tax had been set too high (it was higher than the price paid for the land in some cases), and sales dried up again, contributing to more unrest among the Maoris. In October, the tax was reduced to one penny per acre, boosting sales to the extent that nearly 100,000 acres were soon disposed of, but bringing only a modest amount of revenue into the government coffers, compared with its debts. But by then, the situation with the rebellious Maoris was beginning to get out of hand.

The critical events began with a classic example of the consequences of complacency. FitzRoy had been told before he left London that since the British government had no intention of providing him with troops he should establish a militia in the colony. In a letter dated 11 March, Stanley went further, ordering FitzRoy to introduce proposals for establishing a militia to the Legislative Council. FitzRoy duly did so, in September 1844, at a time when customs duties had been abolished, trade was picking up and the native situation seemed to be improving.

Indeed, things had been so quiet (relatively speaking) that on 16 September FitzRoy was able to complete a task that had been occupying him on and off for the past eight years – he sent back to Beaufort the very last sets of sailing directions for South American waters derived from his work with the *Beagle*. At the same time, in his covering letter (which arrived in London on 14 April 1845) he urges the Hydrographer to send out an officer who can carry out a proper survey of the waters around New Zealand. It had originally been planned that Alexander Usborne, from the *Beagle*, would do this, and he had accompanied FitzRoy as far as Australia before having to abandon the project because of illness, but FitzRoy still had all the required instruments stored at Government House. 'If I were not so hard pressed in *Every* way I would soon teach some body,' he tells Beaufort. But:

the ordinary affairs of this Colony totally engross *my* mind and time. The task I have to grapple with here is beyond anything I could have anticipated in Extreme difficulty: but – thank God – I see no cause for dismay – and have great reason to be thankful for the *perfect* health enjoyed by my wife – our children – her Father and your ever attached friend.

The greatest difficulty of all has been the dishonoring of so many Government Bills – drawn by Mr. Shortland – and the consequent destruction of credit at a period of extreme financial distress. I have been obliged to make a paper currency – and do *many* rash things (as they may be deemed) but – at whatever responsibility – the course I think best for our Queen and country – that I will follow.

Against FitzRoy's advice, the council at this time argued that there was no need for the formal establishment of a militia, and that, indeed, if one were to be established it would antagonise the natives, who would see it as a provocative act. FitzRoy allowed himself to be persuaded; he had obeyed the orders from London in asking the council to consider the proposal, and it wasn't his fault that they had rejected it. In his own communications with Stanley, FitzRoy explained why he had not pressed the case. He pointed out that there was 'so much rancorous feeling' among the settlers towards the natives that it would not be wise to arm them except in case of great emergency, and stressed the need for disciplined troops; these were points soon to be used with great effect by FitzRoy's successor as governor, George Grey, to Stanley's successor, Earl Grey;[26] but they didn't cut much ice with Stanley when FitzRoy made them. Trouble – just the kind of trouble which could have been nipped in the bud by disciplined troops – duly flared up, at the settlement of Kororareka, which had largely shed its 'wild west' image from the days when FitzRoy and Darwin had visited it on the voyage of the *Beagle*, and where some of the wilder young Maori men yearned for the good old days.

The leader of the hotheads was a chief called Hone Heke, who, having initially converted to Christianity but then rejected it, provides an almost too perfect example of what had gone wrong in the northern settlements. Back in July 1844, Hone Heke and a group of his followers had shown their defiance of British authority by cutting down a tall flagstaff at Kororareka from which waved the symbol of that authority, the Union Flag. At one level, the matter was trivial; symbolically, however, it could not be ignored. FitzRoy seemed to do everything right. He restored the flagpole, sent to New South Wales for reinforcements, and in a conciliatory gesture first declared Kororareka itself a free port and then, as he was probably planning to do anyway, abolished customs duties entirely. He reached an understanding with the Christian chiefs, most notably the powerful Waka Nene, that the British would take no punitive action if the Maoris kept Hone Heke and his followers under control in future.[27] And, about the same time, the land sales tax was reduced from ten shillings an acre to a penny an acre. It was the success of these measures that made the settlers feel that the situation had improved so much that they had no need of a militia in September 1844.

Things changed both because the short-lived improvement in the local economy brought about by FitzRoy's measures began to falter,[28] and because rumours spread of a plan by the Colonial Office to impose a tax on unoccupied Maori land, and to claim the land for the Crown if the tax were not paid. In January 1845, Hone Heke cut down the flagstaff again. FitzRoy had it set up, bound with iron, and protected by a permanent guard, part of a small reinforcement of troops that had arrived from New South Wales. At dawn on 11 March, Hone Heke arrived at the head of no more than 200 men, who surprised the guard, cut down the flagstaff, and went on the rampage through the town, causing widespread panic. Women and children were escorted on to the ships in the harbour (HMS *Hazard*, a United States corvette the *St Louis*, the whaling ship *Matilda*, and a schooner called the *Dolphin*), houses were set ablaze, and the small force of soldiers retreated. Amazed by their victory, the Maoris went on what seems to have been a rather restrained looting spree, which encouraged some of the settlers to return to the shore to regain some of their property. According to Thomson, who arrived with the eventual reinforcements from New South Wales, 'one strong-minded woman was seen pulling a blanket against an armed native, and children left on shore in the hurry of the flight were sent by the enemy uninjured to their parents. The whole affair was conducted in the best spirit.' But the township was razed, possibly as much by accident as design.

For the time being, that was the end of the matter. The refugees from Kororareka were taken in at Auckland, where (as if he did not have enough on his mind) FitzRoy's wife was in the late stages of a difficult pregnancy, about to give birth to their fourth child (Katharine, who arrived on 22 March – the fourth of their five babies to survive infancy).[29] Although we know little about FitzRoy's family life in New Zealand, this fact alone stands out as a sharp reminder that concern about a widespread Maori uprising was of more than merely academic interest to him. Happily, though, Waka Nene and the other Christian chiefs remained loyal, while the panic-stricken legislative council rushed through the Militia Bill they had so recently spurned, requiring all male British settlers between the ages of eighteen and sixty to hold themselves ready for service and to be available for military training for up to four weeks each year.

Further reinforcements arrived from New South Wales at the end of March 1845. But they were an unimpressive force some 250 strong, raw recruits who had never been in action, poorly led and with inadequate artillery. With these forces and the friendly natives FitzRoy ordered several attempts to bring Hone Heke and his followers to justice, but the rebel's base in the interior proved impregnable to the forces available, and the result was a stalemate, with Hone Heke holed up in his fortress but with the British, thanks to the reinforcements, now impregnable to any attack from such a small band. But they were still vastly outnumbered by the other Maori tribes. The key factor in the stalemate was that the majority of the chiefs sided with the British, a factor which does credit to FitzRoy's policies, and for which he is seldom praised. By now, the friendly Maoris were about the only people in New Zealand who had any respect for FitzRoy. The settlers, totally disillusioned with the Governor, had already been sending messages to London petitioning Parliament for his recall, and were delighted when news came, on 1 October 1845, that he was, indeed, to go. But in a final irony, it turns out that FitzRoy was not recalled because of those petitions, nor was he really sacked for any of his own failings (although the Colonial Office found enough technical failings on his part to provide an excuse for the recall), but because of the changing political fortunes of the government in London.

FitzRoy's recall was announced in the House of Commons at Westminster on 5 May 1845, before the various petitions from New Zealand reached London, and a few days after the dispatch of the letter (dated 30 April) breaking the news to FitzRoy himself. The official reasons given in the House for his recall were the irregularity and inadequacy of his dispatches, the failure to organise a militia, a lack of firmness regarding 'the native question' and his direct disobedience of instructions in issuing what amounted to paper money and in imposing taxes on the sale of land without authority from London. There was just enough truth in these accusations to provide a smokescreen to conceal the real reasons for his dismissal, but taken individually the charges hardly amount to much. In spite of the difficulties of communication with London, and even allowing for his preoccupation with pressing matters in New Zealand, FitzRoy certainly did not live up to the high standards expected by the Colonial Office in reporting back to them. Towards the end of 1844, he wrote to Stanley offering an explanation and excuse for his tardiness, suggesting that 'a few days' delay would not have been ill bestowed on an honest and hardworking public servant'. On the margin of the letter, Stanley has written 'But I had waited for months!'[30] Even so, this was not, in itself, a sacking offence. As for the militia, it would have been impossible to organise without the active support of the colonists, and as we have seen, that came only when they saw an immediate threat to life and property. The other 'faults' on FitzRoy's part arose directly (as, indeed, did the militia issue) from the failure of the government in London to provide proper financial and military support for the colony – and that is what the sacking was really all about. As Stanley wrote to FitzRoy in May 1845:[31]

> The concern with which I announce this decision is greatly enhanced by the remembrance of the public spirit and disinterestedness with which you assumed this arduous duty, and of the personal sacrifices which you so liberally made on that account; nor can I omit to record that in whatever other respect our confidence in you may have been shaken, Her Majesty's Government retain the most implicit reliance on your personal character, and on your zeal for the Queen's service. You will, therefore, readily believe that I have acted on this occasion in reluctant submission to what I regard as an indispensable public duty.

That 'indispensable public duty' was saving Stanley's own political skin. In 1845, Robert Peel's Tory Government had only a small majority in the House of Commons and was under threat from the opposition; the problems with the New Zealand colony provided a convenient stick with which the opposition could beat the government. According to a contemporary writer,[32] at this time the Company could command forty votes in Parliament, ample (when added to other opposition MPs) to 'have secured a majority against the ministry, if it had not consented to abandon its faithful colonial servant'. It was certainly clear that the affairs of the colony had been mismanaged, and in the spirit of 'the buck stops here' the person who should have carried the can and resigned was Stanley. But Stanley was a career politician who had switched allegiance from the Whigs to the Tories in 1830, and would later (as Earl Derby) serve three times as Prime Minister: as a stopgap in 1852 and then in 1858–9 and 1866–8. Far better (from his point of view) to dismiss FitzRoy, and then carry out the reforms that were required so that an improving situation could be credited to his perspicacity in choosing the new Governor. Nobody in the House was fooled, and when the situation was debated in June speaker after speaker rose to point out that FitzRoy had been put in an impossible situation with, as one Member put it, 'a host of hungry, and, in many instances, useless officials with an inadequate exchequer', and that he had no military support or money with which to carry out the tasks required of him.[33] The ironic outcome of all this was that FitzRoy's successor received both. He also, thanks to a decision made before the identity of that successor had even been decided, received twice the salary FitzRoy had received – a clear sign that London recognised that it had (in much more than economic terms) undervalued the previous incumbents of the post.

George Grey was Stanley's kind of political animal. He had been born in 1812, attended the Army college at Sandhurst and reached the rank of Captain before turning his attention to colonial affairs. He led an exploration of northwestern Australia which resulted in the publication of a two-volume book, *Journals of Two Expeditions of Discovery in North-West and Western Australia*, in 1841 (coincidentally, the small exploring party had been conveyed on the initial stage of its journey to Australia in 1837 by the *Beagle*, then under the command of John Wickham, at the beginning of the voyage to survey the coastline of Australia mentioned in Appendix I). The expedition actually achieved very little, and is puffed up almost out of recognition in Grey's self-serving memoir. By the time it was published, thanks to some careful political manipulation, he had been appointed Governor of South Australia, in 1840 at the age of twenty-eight. By 1845, he had proved a capable administrator, and was more or less on the spot when a replacement for FitzRoy was required in New Zealand. So he got the job. He also got funds with which to put the economy of the colony on a secure footing, and troops with which to bring the rebellious Maoris to heel. FitzRoy would surely have used the troops just as effectively; both in Tierra del Fuego and in New Zealand he had shown that he was not afraid to use force once moral persuasion had failed.[34] He even put into effect FitzRoy's proposal for paying salaries to friendly chiefs, just to make sure they stayed friendly. And he had a stroke of luck when Peel's administration fell in 1846, and was replaced by a Liberal government, under Lord John Russell and with Earl Grey as Secretary of State for War and the Colonies, which further smoothed his path. The result was a reasonably smooth transition for New Zealand from a frontier colony to a fully fledged state[35] in its own right, at the expense of a treatment of the Maori population which was probably harsher than it would have been under FitzRoy, but not as harsh as it might have been if he had never been Governor. Although popular mythology in New Zealand still portrays FitzRoy as a failure, the summing up of historians is that it would have been impossible for anybody to carry out the brief he was given with the resources made available.

It's hard to escape the conclusion that when it came to the Governorship of New Zealand, FitzRoy was the right man, in the right place, at the wrong time – but only by a couple of years. It's not too fanciful to imagine a world in which Grey was appointed in 1843, failed for all the reasons FitzRoy failed, and was replaced by FitzRoy, who succeeded for all the reasons that Grey succeeded in our world. As the *Dictionary of New Zealand Biography* succinctly summarises:

> his determination that the Maori should be treated with fairness and justice, while European settlers should discover their new life in peace and harmony, constituted a major contribution to the life of the new colony. That he had less ostensible success as governor was the result of Colonial Office policy rather than of his own shortcomings.[36]

Thomson's conclusion was that 'Captain Fitzroy's bankrupt finances brought large grants of money, and the destruction of Kororareka large bodies of troops.' Quoting with approval a speech made by the Bishop of New Zealand, Dr Selwyn, at a farewell dinner given for George Grey in 1853, he concurs with the description of FitzRoy as 'the man who lost Kororareka, but saved New Zealand'. The references to FitzRoy in that speech are, indeed, worth quoting in full, bearing in mind that it was made at a dinner to honour his successor, and less than ten years after FitzRoy left New Zealand:

> Next came a man whom I can never think of without sorrow and respect. For, mark me, gentlemen, I cannot measure merit by

success. A good man struggling with the storms of fate will command my sympathy, even more than one standing on the pinnacle of success. I honoured Captain FitzRoy in his misfortunes, as I honour you, Sir George Grey, in your prosperity. Shortly after his recall, I saw a letter from the Secretary for the Colonies, in which he said: 'No one here dares say a word for poor FitzRoy.' I am thankful to have the opportunity of saying a word in New Zealand, which no one would say in England. I have seen that honourable man, for the sake of the public good, sacrificing his own private property, and, what is even dearer to us all, his public reputation. There are many here present who can recollect the time, so different from the present, when this colony was on the verge of bankruptcy. I have seen my honourable friend, the Colonial Treasurer, who now sits so comfortably upon his well-filled chest, reduced almost to despair; and I have seen my trusty friend Mr. Kennedy, not then, as now, the officer of the Union Bank of Australia, and helping to pay the proprietary a dividend of forty per cent. – not then, as now, rejoicing in deposits to the amount of 120,000*l.*, but preparing at four o'clock in the afternoon to close the doors which, at ten o'clock the next morning, he had resolved not to open to the public. Then, in the face of his instructions, at the risk of loss of office, with no possible advantage to himself, right or wrong according to political economy, well or ill as to the result, for the sake of the public credit, and for no other cause, Captain FitzRoy made debentures a legal tender, and lost his office in his attempt to save the colony from ruin. One instance more, gentlemen, – for I shall not weary you with many: I was with Captain FitzRoy at the meeting with the native chiefs of the north, when the reinforcement of 200 men arrived from Sydney. On that occasion Thomas Walker [a friendly chief] gave the pledge, which he has since amply redeemed, that if the Governor would give the lie to Heke's assertions, that the land was to be taken from them, by sending back the troops, he and his men would guarantee the protection of the north. It was the wisest as well as the bravest act that was ever done by any Governor in the British Empire. It is true that the native allies were a little too late in taking the field; but when they took it they kept it. The example was set, and from north to south no British force has ever been employed without its contingent of native allies. The effect of that alliance, it is for military men to estimate rather than for me. I simply state the fact; and if I were to write the history of Captain FitzRoy's administration, it would be in these words: 'He was the man who lost Kororareka, but who saved New Zealand.'[37]

Dr Selwyn was not alone in this assessment. The archive of the Dixson Library in New South Wales includes the draft of an article, or memorandum, written by the geographer Trelawney Saunders (undated, but from internal evidence written in the mid-1850s, in response to a publication giving all the credit for the success in New Zealand to Grey) in which he says that FitzRoy 'was the saviour of New Zealand; and the most critical events in the history of the colony occurred during his administration'. After emphasising how much FitzRoy gave up in England to go to New Zealand, and referring to FitzRoy's 'high sense of public duty', Saunders offers the opinion that if he had followed the letter of his instructions and waited for approval from England before taking the steps he did,

he would inevitably have permitted the utter ruin of 10,000 English settlers, and have risked the loss of the colony. He preferred the exposure of his own repute and fortune. His private property was sacrificed, and his reputation became the scape-goat for the mercenary & blood-guilty malpractices of the New Zealand Company, and for the traditional do-nothing indifference of the Colonial Office.

Saunders sums up his reasons for writing the paper, saying that he will be satisfied 'if the credit due to a honourable man, – who has done good and disinterested service to his country, under most arduous and critical circumstances, – has been maintained against the oversight of a contemporary writer, and raised above the malices which he has happily outlived'. That description – *good and disinterested service to his country, under most arduous and critical circumstances* – could be a summary of FitzRoy's life.

We can only hope that FitzRoy learned of the sentiments expressed by Selwyn and by Saunders. Although not one to express any bitterness he felt in public, it must have been galling for him to see the success Grey achieved (including an eventual knighthood) with the aid of exactly the kind of support that FitzRoy had requested but been refused. And FitzRoy's mental equilibrium was possibly not helped by the coincidence that his half-brother Charles was appointed Governor of New South Wales in the same year that FitzRoy was recalled from New Zealand. Charles, too, received a knighthood for his service to the Crown. One curious side-effect of all this is that there are two Fitzroy Rivers in Australia. The one in Queensland is named after Charles FitzRoy, while the one in Western Australia was named by Lieutenant John Lort Stokes, on his surveying voyage in command of the *Beagle*, in honour of his old Captain, Robert FitzRoy.[38]

FitzRoy handed over to Grey in November 1845, ensuring a smooth transition by providing all the help and advice his successor needed to become established, and set off home with his family in January 1846, on a ship bound eastward around Cape Horn, completing his second circumnavigation of the globe (this time from west to east, the other way round from the *Beagle* voyage), and passing close by the scenes of some of his triumphs with the *Beagle*. In his reply to Stanley following the news of his recall, he wrote[39] of his conviction that he was 'deeply and irreparably injured' by the nature of the dismissal; but from a distance he cannot have been aware of the strength of support for him in the House of Commons; in fact his prospects and reputation in England seem scarcely to have been affected by the unfortunate antipodean interlude, and he bounced back with a vengeance once he had got settled again back home.

CHAPTER EIGHT

Unrequited Hopes

ROM 1846, WHEN FitzRoy returned to England with his family, to 1854, when he became the first head of the Meteorological Office, we know everything and nothing about his life. We have the dates, names and places from the official records and formal letters, but there is very little to flesh out these bare bones, in the absence of a Darwin or a Sulivan working alongside the Captain and keeping his own record. But from what we know of FitzRoy (not least, his usually unbounded optimism punctuated by occasional periods of deep depression) it is often possible to read between the lines.

To set the context of his – this time – less than happy return, we can take stock of his family situation in the summer of 1846, when he was forty-one years old, an unemployed ex-Governor of New Zealand. There were now four children below the age of ten (born in 1837, 1839, 1842 and 1845), his wife, and his father-in-law (now in his sixties). He had some money, but by the standards of the FitzRoy family he was essentially broke and, quite apart from his own restless need to be active and doing good, he needed a job in order to provide income,[1] even though Major General O'Brien must have been contributing something while he was a member of the household. Typically, FitzRoy didn't dwell on the past, but looked forward, to his own future and the future of the Royal Navy. What he really wanted was another seagoing command, and he could clearly see that the future of the Navy lay in steam, not sail. Indeed, during his time with Trinity House, between 1838 and 1843, he had often travelled in steamships in British waters, and taken a keen interest in their operation. Now, learning that in 1847 Sir Charles Napier was to take out a small fleet of ships of various kinds on an experimental cruise to test and compare various technologies, he obtained a place for himself as a passenger and observer in the little squadron, first on the *Vengeance* and then on board a little steamer driven by a screw, rather than paddles – the *Amphion*. This was still daringly new technology for the Navy, although Isambard Kingdom Brunel's screw-driven iron ship *Great Britain* had been launched in 1843.

It's worth emphasising that in spite of his difficulties in New Zealand, FitzRoy was far from being in disgrace on his return home, and there was widespread recognition, as we have mentioned, that he had been made something of a scapegoat for the failings of others (in this connection, it didn't do him any harm that there had been a change of administration in 1846). Indeed, for a couple of years after his return to England, he was regularly consulted by the Colonial Office about the dispatches being received from his successor, George Grey.

We don't know what strings (if any) were pulled, but FitzRoy's keen interest in new technology paid off in July 1848, when he was given the task of overseeing the outfitting of a new frigate, HMS *Arrogant*, which was essentially a sailing ship but with an engine as well – a steam engine driving a screw propeller. She would be the Navy's first screw-driven steamship, a vessel of 360 tons and thirty-six guns, with an engine producing just 360 horsepower (that's a little more power than an MGZ sports car, which weighs only 1.5 tons, but a lot less than, say, the 50,000 horsepower of the *Titanic*). There were many other technical innovations in the ship, and FitzRoy was responsible for making sure everything was installed properly and the fitting out proceeded smoothly; while he was doing so, for a couple of months in the autumn of 1848 he was also Acting Superintendent of the entire Woolwich Dockyard.

It had always been understood that FitzRoy's task would not end when the ship was ready for sea. All of the innovations had to be tested and refined during sea trials, and when the *Arrogant* was officially commissioned in March 1849, Robert FitzRoy was appointed as her Captain. For the rest of the year, she seems to have spent most of her time in home waters, making frequent visits to the dockyards at Woolwich and Portsmouth for adjustments and improvements to the new machinery. But when everything was to his satisfaction, the ship was dispatched to Lisbon – a proper seagoing command, with the independence (as Captain of his own ship) he always seemed to crave. According to the *Good Words* obituary: 'He commanded her with great success; and that same happy tact and consideration for others, which made the "Beagle" a home to all in her, caused him to be equally popular in the "Arrogant".' As the icing on the cake, FitzRoy was accompanied on the ship by his son, Robert O'Brien FitzRoy, who had been born on 2 April 1839 and at the age of ten was considered old enough to begin following in his father's footsteps.

But the happiness was not to last, and once again FitzRoy was to have his hopes dashed. Our best insight into what happened in Lisbon comes from our knowledge of FitzRoy's breakdown on the west coast of South America, and a curriculum vitae which FitzRoy wrote in March 1852, while seeking employment (we do not know the specific purpose it was written for, but a copy survives in the Alexander Turnbull Library in Wellington).[2] The first relevant factor is that for the best part of two years FitzRoy had been under intense pressure, driving himself in his usual perfectionist way to get the *Arrogant* fully seaworthy and fit to take her place in the fleet. FitzRoy would not have wanted his ship to be a mere novelty or curiosity, or even to be accepted as an ordinary member of the fleet; he would have wanted her to be the best frigate in the British Navy (which, to him, would mean the best in the world) regardless of whether she was powered by steam, sail, or a combination of the two. Even without detailed accounts of his day-to-day activities during this time, we can be sure that the former Captain of the *Beagle* had lost none of his tautness in the intervening years, and that plenty of hot coffee (along with tact and understanding) was poured out during the process of getting his ship up to the mark.

In Lisbon, another factor stopped FitzRoy from sitting back and enjoying his command. He was the senior Captain among the British ships assembled in this friendly port, and as such was expected to do his part in showing the flag by entertaining on a suitable scale – a 'suitable scale' involving far more expense than a Captain's pay could support, and making further inroads into his already depleted resources. The last straw seems to have been news from home with what turned out to be exaggerated accounts of the financial problems his wife was experiencing – but there was no way for him to know that these difficulties were not as bad as he heard.

The combination of factors seems to have flung him into a deep depression very similar to the one that he was brought out of by the combined efforts of Wickham and Bynoe back in Valparaiso in 1834. But this time, with a relatively new group of officers under his command – men who did not know him as well as Wickham, Bynoe and the other officers on the *Beagle* had known him, and who might be expected to be more interested in their own promotion prospects than Wickham had been – and with England so close to hand, things took a different turn. In that c.v., FitzRoy simply says:

In February, 1850, after having proved the Arrogant in every way – and fairly tired himself out – Captain FitzRoy was obliged to yield to the effects of fatigue – and anxiety about home affairs – conjoined; which had unnerved him, for a time.

At Lisbon, he consulted with Commodore Martin, and gave up his ship, in order that he might settle his domestic affairs, and regain his usual uninterrupted health. A week's change of air only, with absolute rest, sufficed to make him feel himself a different person: and a few months in England, after arranging difficulties that had harassed him, entirely recruited his health, which since that time has been, as it always was throughout his whole previous life – remarkably good.

The purpose of this memorandum (which conveniently glosses over the incident in Valparaiso!) seems to have been to assist FitzRoy's efforts to obtain another ship. Once he had sorted out his domestic affairs and been restored to health, he had taken a position as one of the Managing Directors of the 'General Screw Steam Shipping Company', which occupied him for a year (until late 1851) in the affairs of a company closely involved in the technology that was now clearly at the heart of future naval ship design. This was obviously a position he had sought because of its relevance to his career in the Navy, and after his year in office (the limit allowed by the rules of that company) 'his sole object now is to follow up his own proper profession as soon as he can obtain employment in command of a ship. While engaged with that Steam Company he had opportunities of collecting much information of various kinds that will be useful to his own Profession: the interest of which has ever been uppermost in his mind.'

But there was no way back. There were many more Captains on the Active List of the Royal Navy competent for command than there were ships for them to command in those relatively peaceful times, and the view of the Admiralty was (quite rightly) that a man who had resigned his command once on grounds of ill health might do so again, and possibly in more awkward circumstances than when at anchor in a friendly port.[3] There had been one consolation. In 1851 (the year of the Great Exhibition held in the Crystal Palace at Hyde Park), FitzRoy had been elected as a Fellow of the Royal Society. The formal proposal for his election, giving his address as Norland Square, Bayswater, was supported by thirteen Fellows including not only Beaufort and Darwin but the naturalist Richard Owen and Charles Wheatstone (of whom more later). The achievements picked out by the election notice as qualifying FitzRoy for Fellowship were his authorship of the *Narrative*, the invention of a new kind of surveying quadrant and the fact that he was 'distinguished for his acquaintance with the science of Hydrography and Nautical Astronomy', and described him as 'eminent as a Scientific Navigator, & for his chronometric measurements of a chain of meridian distances during the circumnavigation of the globe which he conducted'. He must have been genuinely puzzled at the Admiralty's failure to offer him another ship, with such impeccable credentials and convinced, as he would have been in his own mind, that there was no prospect of him ever suffering another day's illness in his life. But before he could pursue the matter further, and perhaps bring the full weight of the FitzRoy influence to bear in a last attempt to get a ship, all his plans were thrown into confusion, and he had to focus entirely on domestic affairs for the next two years.

The optimistic c.v. we have quoted from is dated 15 March 1852, and there is no hint of any problems at home which might stop FitzRoy from taking up a seagoing command. But on 5 April, less than three weeks later, his wife Mary died. The news comes out of the blue, and there is no hint of any preceding illness; she was only in her fortieth year, and was survived by her father (who lived until 1855) and by her four children, the eldest not yet fifteen. The only details we have come from two undated letters from FitzRoy to his sister Fanny:

Monday my wife got a chill – when out – that struck inwards & on Monday night she felt acute pain about the heart – left side and back.

This increasing – I called the Doctor at four on Tuesday morning – In the afternoon of Tuesday she got worse – that night was a bad one – on Wednesday – no better – Yesterday I sent to Locock who saw her in the evening. Last night was *very* trying . . . Today she is *rather* better . . . She is *now* easier and has slept an hour. Do not say anything to Friends.

Fanny and her husband immediately came to help, as the second letter makes clear:

May God bless you and your most kind husband for your unremitting and most tender care of my beloved wife in her heavy affliction. How She would have borne up – but for your kindness – I shrink from thinking.

She is tolerably composed and quiet now – and has had a *little*
sleep . . . I will let you know again bye & bye how she is.

But there was to be no good news.

Nowhere in the FitzRoy story is the lack of any domestic detail
more frustrating than for the years from 1852 to 1854. With FitzRoy's
family, we have the bare facts and dates of key events such as births,
marriages and deaths – but unlike with the Darwin family we have no
details of the troubles of pregnancy, difficulty of labour, romantic (or
unromantic) courtships, or terminal illnesses. Obviously, FitzRoy's
attention was focused on his family over the two years following the
death of his wife.[4] Robert, now thirteen, was old enough to stay with
his father (he joined the Navy in January 1853), and at least in the short
term the two younger girls, Fanny (not quite ten) and Katherine (just
seven) could stay with the family of George FitzRoy, Robert's brother.
Fourteen-year-old Emily was more of a problem. The problem seemed
to be resolved almost immediately by a generous offer from Lady Vane
Londonderry, the wife of Robert's maternal uncle, the Marquess of
Londonderry. She was the heiress, you will recall, who brought the
Durham connection and coal-mining wealth into that branch of the
family. Now in her early fifties (her husband, Robert's uncle Charles,
was in his seventies, and would die in 1854) and with no young children
of her own to worry about (although she had had three boys and two
girls of her own), she offered to take Emily into her own household.
The intentions were undoubtedly good; Lady Londonderry wrote to
FitzRoy, just two weeks after his wife had died, that:

> With your eldest of 14 I do hope, that with the Maternal feelings
> I should at once extend to her, she would be happy – while I should
> feel conscious of fulfilling a sacred charge which I have
> spontaneously offered to the Grand Daughter of my Husband's
> eldest and dearest Sister.

Some idea of the kind of circles Lady Londonderry moved in, and the
kind of future there might have been for Emily if things had worked
out differently, can be gleaned from the fact that Lady Londonderry's

own daughter, Frances Anne Emily Vane (who had been born in 1822 and was Robert FitzRoy's first cousin) married John Winston Spencer Churchill, the Duke of Marlborough, and became the mother of Randolph Churchill, himself the father of Winston Spencer Churchill.[5] But although the arrangement seemed promising in theory, it did not work out in practice.

Emily seems to have inherited many of the characteristics of both her father and his sister Fanny, keeping her feelings to herself but being very hardworking and diligent. In a formal agreement setting out her terms for looking after the girl, Lady Londonderry spelled out many restrictions, including a requirement that all her letters (incoming and outgoing) except those to and from her father were to be read; it would, indeed, be a strict Victorian upbringing. Little allowance seems to have been made for the fact that Emily had just lost her mother and been removed from the bosom of her family, and while Lady Londonderry continued her grand life, travelling from house to house and busy with social engagements, the girl was left for most of the time in the care of a governess who, it turned out, knew less than Emily did about many subjects she was supposed to be teaching and was no musician, able only to listen when Emily played the piano (which she did rather well) rather than provide any tuition at all. When Lady Londonderry read correspondence between Emily and her father (against the terms of her own 'contract') which hinted at Emily's unhappiness, this led to an exchange of letters[6] which culminated in November 1852 in FitzRoy's decision to move Emily into the care of her mother's sister, pending a more permanent home. Fanny's husband, who had succeeded to the title Baron Dynevor on 9 April 1852, acted as an intermediary in these rather delicate negotiations, which seem to have been accomplished reasonably tactfully.

In correspondence relating to these arrangements, dated 5 November, FitzRoy mentions to Dynevor how far advanced plans were for the tidal survey we have referred to, and how this would mean finding permanent homes for all the girls, although Robert would be able to accompany him: '*Last week* I was *asked* by the Hydrographer to take the command of it – and I have now consented. *When* it will be ordered *officially*, and commences, is not *yet* settled – but it *will* be in the course of a *few* months.' But it never was, largely on financial grounds, and with it went FitzRoy's last chance of a seagoing appointment.[7] It was in the wake of this disappointment, and the troubles involving Emily (though not her own fault), that he decided (as many a Victorian widower with a young family would) that he ought to marry again in order to provide his children with a mother. He chose a cousin, Maria Isabella Smyth, in a union that seems to have been based on affection and practicality rather than being a grand passion. The wedding took place on 22 February 1854; we don't know Maria's age at the time, but her parents had married in 1814 so she was probably no more than thirty to FitzRoy's forty-eight.

Maria was actually rather more closely related to Robert than the simple expression 'cousin' conveys, and was a FitzRoy in all but name, since not only her mother but her grandmother on her father's side was also a FitzRoy. Georgiana, one of the daughters of Augustus Henry FitzRoy (the third Duke of Grafton, and Robert FitzRoy's grandfather), had married John Smyth in 1778. Their son, John Henry Smyth, had married Elizabeth Anne FitzRoy, herself the daughter of George Henry FitzRoy and therefore (like Robert FitzRoy) a grandchild of Augustus Henry FitzRoy. They were the parents of Maria. Robert and Maria FitzRoy had one child together, a daughter named Laura, who was born on 24 January 1858, and would live until 6 December 1943, dying just six months after the last surviving child of Charles and Emma Darwin, their son Leonard.

It is no coincidence that FitzRoy returns to a fully active public life (and therefore becomes more visible to the historian) following his second marriage and the consequent easing of his domestic responsibilities.[8] He must by now have realised that there was little hope of a naval command for a Captain without a ship for four years and coming up to his fiftieth birthday, but there had already been dramatic developments in the Black Sea region, and slightly less dramatic developments elsewhere, which opened up new possibilities, and would soon lead him in an unexpected direction. In 1852 the Duke of Wellington, who had been Commander in Chief of the British Army for a quarter of a century, died. His successor was Henry Hardinge, a career soldier who had been born in 1785 (making him about the same age as FitzRoy's father-in-law) and served with distinction in the Napoleonic Wars (losing his left hand in the process), as an MP (including, in the 1820s, a spell as member for Durham), as Irish Secretary and Secretary of War. He was later (from 1844 to 1848) Governor-General of India, and would be elevated to the peerage, as Viscount Hardinge of Lahore and Durham in 1855. It was Hardinge who, at the urging of Prince Albert, established the first permanent training camp for the British Army, at Aldershot in Hampshire. And in 1821 he had married Emily Jane Stewart, the sister of Frances Anne Stewart, FitzRoy's mother.

By February 1854, at the time of FitzRoy's marriage to Maria Smyth, it was clear that Britain would soon be involved in the conflict brewing in the Black Sea between Turkey and Russia; the British and French fleets had entered the Black Sea to protect Turkish shipping on 3 January 1854, and war on Russia was actually declared on 28 March. It was also clear to those who moved in the right circles, as we shall see in the next chapter, that a job almost tailor-made for FitzRoy was likely to become available in the summer. But in the short term, FitzRoy was still unemployed, and his maternal uncle needed a private secretary to help with the additional workload of the Commander in Chief at the beginning of what became known as the Crimean War. It was not a job that FitzRoy would have welcomed on a permanent basis – although his background as a naval officer, an MP and Governor of New Zealand gave him impeccable credentials for the post even without the family connection, he knew only too well from his time in New Zealand how frustrating it would be for him to be involved for long in such a task. But on the clear understanding[9] that he was available as a stopgap, to help Hardinge move his office on to a war footing, and that he would be released from these duties if and when the anticipated offer came his way, he agreed to help out. It was, after all, his duty to do so. But we can easily imagine the relief he must have felt, after all the frustrations and disappointments, both professional and personal in the years since he had resigned his command of the *Arrogant*, when that offer did materialise and he was at last able to achieve something on land to rank with his achievements in command of the *Beagle*. There would be more frustrations along the way, and it would end unhappily eleven years later. But, in both his professional and his private life, 1854 marked a new beginning for Robert FitzRoy, and during those eleven years he made an enduring mark on the world. To put that new beginning in context, we have to take a step back in time and away from the Black Sea.

CHAPTER NINE

Prophet without Honour

RUSSIA'S INVASION OF the Balkans and attack on the Turkish fleet, which led to the Crimean War, were not the only historically important events of 1853, although they tended to overshadow everything else. One event which seems far more significant in hindsight than it appeared at the time, and which was to shape the rest of FitzRoy's life, was a conference in Brussels, in the last week of August and the first week of September that year, where delegates from ten nations met to discuss the possibility of establishing a uniform system of meteorological observations at sea, and sharing information about the weather. The British were unenthusiastic participants in the conference, which was inspired by the efforts of an American naval officer, Lieutenant Matthew Fontaine Maury, who was then superintendent of the US Navy's Depot of Charts and Instruments, having previously been involved in surveys of the harbours of the southeastern United States. Maury had been born in Fredricksburg, Virginia, in 1806, and joined the US Navy in 1825; but although it was his initiative that prodded the European governments into starting to take meteorology seriously, his own interest in the weather had been stimulated by an Englishman, John Fox Burgoyne (son of the John Burgoyne who had been a British General in the American War of Independence), who had initially suggested a collaboration between the British and the Americans alone.

The time was certainly ripe for some such collaboration, whoever was to be involved. As is so often the case in science, progress depended crucially on the development of appropriate technology, and in this case the standard instruments of meteorology, the thermometer and barometer in particular, were only invented in the seventeenth century and developed for widespread use in the eighteenth century. As more widespread records were kept for longer intervals of time, it became clear that the weather did not operate in a capricious way, but that there were distinct patterns which affected wide areas in the same way, so that it might be possible to develop an understanding of how wind flow worked, with obvious benefits to a world where international trade, communications and warfare depended so much on sailing ships. By the 1830s, there were two rival schools of thought about the pattern of wind flow associated with cyclonic storms, a particular hazard to mariners. One held that the winds blow around the centre of the storm (anticlockwise in the northern hemisphere, clockwise in the south); the other held that the winds blow straight into the centre of the storm, where the atmospheric pressure is lowest.[1] The first idea was put forward by an American steamboat engineer, William Redfield, on the basis of a study of hurricanes and other severe storms. It seemed reasonable to suppose that the same pattern of circulation applied to the lesser cyclonic systems, known simply as depressions, that plague the North Atlantic and (we now know, although Redfield did not) track across that ocean from west to east. Redfield's ideas were taken up by a British Army officer, William Reid, of the Royal Engineers. He was sent to Barbados in 1832 to oversee the reconstruction of government buildings wrecked in a hurricane, and became interested in finding out more about such storms.

Reid's ideas were developed in an extended correspondence with Redfield, and in 1838 he published a set of rules which would help sailors, by observing changes in wind speed and direction, to avoid storms at sea. He persuaded the British government (the same Colonial Office responsible for the fiasco of FitzRoy's time in New Zealand) to start obtaining meteorological observations from its colonies around the world, and then suggested to Burgoyne, the Inspector-General of Fortifications of the British Army, that the Royal Engineers might set up another network of observing stations. The proposal was put into practice in 1851, and it was then that Burgoyne, with official approval, made contact with the US government to suggest extending the network to include the Americans. Burgoyne's proposal was passed down the line to Maury, and Maury came back with the idea of the international conference, which Burgoyne, who would have preferred a cosy arrangement between the English-speaking nations, referred to the Royal Society. The Royal Society supported Maury's proposal for a conference to discuss establishing a uniform system of observations at sea, but held back from endorsing a similar collaboration on land on the grounds that there were so many different systems already in operation.[2]

So the conference took place, with the British delegation (headed by F. W. Beechey) under strict instructions not to commit the government to any expenditure. It determined unanimously a standard system of making observations of the weather at sea and of recording them in a log, and in due course the British decided to spend a little money on the project after all, setting up a new Meteorological Department (or Office)[3] within the Marine Department of the Board of Trade (the Marine Department itself, with Beechey as its head, or Professional Member, had been established as recently as 1850, when safety at sea began to be taken seriously at government level). The formal vote for the funding of the new office was passed by the House of Commons on 30 June 1854, when the Member for Carlow, Mr. J. Ball, stood up to urge that:

The observations made . . . upon land as well as at sea would be collected, as, if that were done, he anticipated that in a few years, notwithstanding the variable climate of this country, we might know in this metropolis the condition of the weather 24 hours beforehand.

Hansard records the response of the House:

Laughter.

Robert FitzRoy had already been lined up for the job of running the new Meteorological Office even before this vote was taken. Lord John Wrottesley was an aristocrat who had been born in 1798 and inherited his ancient title and the family estate at Wrottesley, in Staffordshire, in 1841. Like several of his peers, he was interested in science (although no great shakes as a scientist himself) and he became a prominent Fellow of the Royal Society who would soon (in 1854) become its President; he was also a founder member of the Royal Astronomical Society. Wrottesley had played a significant part in persuading the British government to attend the Brussels Conference, and in the interval between the conference and the official decision to implement its proposals, he had asked FitzRoy, known as something of an expert on weather at sea, about the practical details of carrying out the work proposed in Brussels.

FitzRoy's response, a detailed memorandum setting out just how Maury's ideas might be put into practice (and, incidentally, referring to charts which would provide 'a synoptic view' of the weather), was dated 3 February 1854, and is preserved in the Public Record Office. Part of his advice is to 'employ some nautical man who is interested in such subjects – whose character will guarantee his proceedings, and who will give full time and thought to their pursuing'. It was clear who he meant, and there was more than a hint from the correspondence that if such an office were established there would be a job for FitzRoy if he wanted it. FitzRoy let it be known that if an offer came it would indeed be welcome – which is why he accepted the post with Hardinge only on a pro tem basis. The reasons why he accepted the offer, when it was duly made, seem clear to anyone who has studied his character, even though at a more casual glance it might be thought that such a poorly paid post, with no prestige attached to it, might be beneath the dignity of a former Governor of New Zealand. FitzRoy accepted the post for the same reason that he had accepted that governorship – without a thought for himself, but out of a sense of duty, seeing an opportunity to do good for others. In this case, the others he might do good for were sailors at risk from storms at sea. There is no way Robert FitzRoy could have refused such a post.

FitzRoy took up his new appointment on 1 August 1854, at the age of forty-nine, in the same year that he had married his second wife, Maria; it was clearly his last chance to make his mark in the world. His official title was Meteorological Statist (we would now say Statistician), reflecting the fact that his official brief was to collect and collate data from ships at sea, building up a database which could be used to develop an understanding of weather patterns. His initial salary was a modest £600 per annum, half paid by the Board of Trade and half by the Admiralty; the total budget of his Office in the first year was £4,200. The underlying justification for the cost of this work was that it would enable ship owners and their captains to make more efficient use of their vessels, saving time and money; inevitably, though, FitzRoy was more concerned with the possibility of saving lives, as he had been in his work as an MP, and this would soon lead him to exceed his brief.

At first, everything went as the holders of the purse-strings had planned. FitzRoy had a staff of only three clerks to assist him (there should have been four, but the one appointed to be FitzRoy's deputy was not allowed to move, being regarded as too valuable to let go by the Statistical Department of the Board of Trade) and a modest amount of office space in a building in Parliament Street, close to the present-day location of the Cenotaph.[4] As Burton points out, this was all part of a growing interest in meteorology in Britain at that time. The first full-time government meteorologist, James Glaisher, had been appointed as Superintendent of the Magnetic and Meteorological Department at the Greenwich Observatory in 1840. Glaisher was a 'hands-on' meteorologist, famous for a series of twenty-nine balloon ascents that he carried out between 1862 and 1866 to measure how meteorological conditions change with altitude. The British Meteorological Society (now the Royal Meteorological Society) had been established in 1850, with Glaisher as its most prominent founder member. And there was also the Kew Royal Observatory at Richmond (since 1841 under the control of the British Association for the Advancement of Science, or BA) which carried out its own meteorological and magnetic work. It was at Kew that the instruments required by FitzRoy's Office were tested and standardised before being sent out to the ships which made the observations for him.

The system worked through a network of paid agents which FitzRoy employed in the ports around the British Isles. They found ships' captains who were willing to take part in the data gathering, and supplied them with the standardised instruments (chiefly the barometer and thermometer) either on loan or at cost price; the agents themselves, who collected the weather logs when the ships returned to port and passed them back to London, received 50 shillings for each vessel 'serviced' in this way. The captains were also given charts and other information based on Maury's work, and provided by the US government; but although a satisfactory working relationship with the Americans was established, many on the British side, including FitzRoy, had personal objections to the way, as they saw it, Maury went in for self-promotion and sought publicity. This was, of course, the very antithesis of FitzRoy's approach to public work as a matter of duty for the common good. In a letter to John Herschel, written on 4 May 1858, FitzRoy stops just short of directly accusing Maury of plagiarism:

He has collected facts (or data *some* doubtful) very industriously (aided by a large staff) and has set other people to do likewise. He has given good sailing Directions – and has duly trumpeted – according to the fashion (however unworthy) of the day: therefore – in America – he has a large reputation among men of my cloth – who have not heard quite so much of old Dampier – Cook – Flinders – Dalrymple &c &c – as *educated* men in England have generally. Maury's adoption of these men's ideas – and non-recognition of their origin – is sad.

And on 2 May 1865, writing to Charles Darwin, the naturalist Joseph Hooker includes both Maury and Glaisher in a condemnation of 'those cattle who live by self-glorification'.[5] But this did not affect the work of the Meteorological Office, which had supplied fifty merchant ships and thirty Royal Navy vessels with the required instruments by May 1855.[6]

But FitzRoy was not someone who would be prepared to while away his time in Parliament Street waiting for the data to come in. With the system up and running, he was free to make the best use he thought appropriate of his time. Although it was intended that the Office should have a primarily (if not entirely) statistical role, there was nobody to check up on FitzRoy, and his guidelines for the running of the Office, such as they were, were contained in a letter from the Royal Society to the Board of Trade, which was not even written until 22 February 1855, by which time FitzRoy had had his feet under the desk for six months. He chose to regard this as advisory rather than containing his orders – and FitzRoy was good at ignoring unwelcome advice. The letter (which is reproduced in full as one of the Appendices to FitzRoy's *Weather Book*) goes into tedious detail about the techniques to be used in making observations of the weather and keeping records, but one short passage gave FitzRoy a justification, at least in his own mind, for much of his later work:

It is much to be desired, both for the purposes of navigation and for those of general science, that the captains of Her Majesty's ships and masters of merchant vessels should be correctly and thoroughly instructed in the methods of distinguishing *in all cases* between the rotatory storms or gales, which are properly called *cyclones*, and the gales of a more ordinary character, but which are frequently accompanied by a veering of the wind, which under certain circumstances might easily be confounded with the phenomena of cyclones, though due to a very different cause. It is recommended, therefore, that the instructions proposed to be given to ships supplied with meteorological instruments should contain clear and simple directions for distinguishing *in all cases* and *under all circumstances* between these two kinds of storms; and that the forms to be issued for recording the meteorological phenomena during great atmospheric disturbances should comprehend a notice of all the particulars which are required for forming a correct judgement in this respect.

In readiness for the anticipated flow of data, FitzRoy first devised a new diagrammatic system for recording weather observations, based on Maury's system but improving on it. The diagrams, which FitzRoy called 'wind stars', each covered a square of ocean 10 degrees on a side, and contained weather information for one of four three-month periods (January–March, April–June, July–September, October–December), so that (once all the data were in) a seafarer would be able to tell at a glance what kind of winds to expect in a certain patch of the ocean at a certain time of year. While the data were coming in and the wind stars (which would indeed prove a great boon to seafarers) were being compiled, FitzRoy, drawing on all his experience of making observations in gales around Cape Horn from the heaving deck of the tiny *Beagle*, designed a particularly rugged, reliable and simple barometer, which became a standard issue, and wrote a manual with instructions for its use in foretelling weather changes that was first published in 1858 and was reprinted in new editions many times. He also served on a committee investigating the way the deviation of a compass is caused by the presence of iron in ships, and how the problem might be overcome.

But while all this was going on, one particular weather event triggered the application of new technology to meteorology. On 14 November 1854, during the Crimean War (which ended in 1856), a combined Anglo-French fleet off Balaclava was struck by a severe storm which caused considerable damage to the ships. Long after the event, the French astronomer Urbain le Verrier, who also had an interest in meteorology and organised the first French meteorological service,[7] was studying weather records of the relevant weeks sent back to Paris from observing sites across Europe, and realised that the storm had travelled from west to east across the continent, so that in principle it would have been possible to send a warning, using the relatively new electric telegraph, to the Crimea. We now know that this west to east movement of the atmosphere is characteristic of the latitudes of Europe and North America; FitzRoy was one of the first people (see *The*

Weather Book) to realise that this eastward movement of the air is a large-scale feature of the workings of the weather machine.[8] As le Verrier realised, the new technology was just becoming widespread enough to make such warnings practicable in the future.

The electric telegraph had been invented by William Fothergill Cooke and Charles Wheatstone in England, who took out key patents in 1837 and 1845; independently, and knowing nothing of their work, in America Samuel Morse hit on the same idea, in the 1830s, and (with his partner Alfred Vail) developed it much further, into a practical system of communication (including his eponymous code, invented in 1838) in the 1840s. By the middle of the 1840s, meteorological observations from far and wide were being gathered using the new technology in both Europe and North America. Realising that at least some storms travel from west to east across the North American continent, as early as 1847 Joseph Henry, Secretary of the Smithsonian Institution, proposed a network of telegraphic links to give warnings to citizens of the eastern States of storms coming from the west. By 1849, more than 200 widely scattered observing stations were reporting to the Smithsonian in Washington, where weather information – that is, information about what the weather had been like across the US at the time the reports came in – was displayed for the public on a large map. Daily weather reports (but, crucially, *not* forecasts) were also published in the *Washington Evening Post*, until the system was disrupted by the American Civil War (1861–5), which allowed the Europeans to leapfrog ahead.

In London, the *Daily News* commissioned James Glaisher to collect weather data from a network of observing stations around Britain, and published its first 'Daily Weather Report' (again, not a forecast) on 31 August 1848. The reports were entirely text, with no charts; but in 1851 Glaisher produced a series of daily weather charts, based on the same information, which, although not published in the newspaper, were put on display for the edification of visitors at the Great Exhibition. Thanks to le Verrier's initiative following the Crimean storm at the end of 1854, the French took the lead in Europe in setting up an official network of telegraphic weather warnings; but FitzRoy had barely got started at the Meteorological Office at this time, and, as we have seen, was preoccupied initially with establishing the system for collecting and collating data from ships at sea. Besides, although as a naval man he was concerned about what happened to the fleet off Balaclava, his brief was specifically to gather data from the oceans, not to worry about the weather over land. But as the routine work of the Office began to function smoothly, FitzRoy was able to devote an increasing amount of his time and energy into looking at the broader picture, and in particular to finding ways of giving warnings to mariners of imminent storms at sea. He had had something of a bee in his bonnet about this since the nearly catastrophic encounter of the *Beagle* with a pampero soon after he had taken command. Through careful attention to the way the barometer had responded to changing weather conditions during his years at sea, he had worked out a series of rules of thumb by which the way the barometer changed could be used to foretell changes in the weather. As early as 1843, shortly before he left Parliament to take up the Governorship of New Zealand, he had outlined, in evidence to a parliamentary select committee investigating shipwrecks, a proposal to use this knowledge to set up a network of barometers at observing stations around the coasts of Britain, so they could be used to provide warnings of storms to sailors. Now, he was in a position to put this idea into practice, with the bonus of the growing mass of data on weather patterns that he was now accumulating. With these data, he could construct charts which showed the pattern of weather over a wide area of ocean on a single day; the only snag was that since the data all came from ships, and had to be physically handed to the agents in the various ports and then conveyed to London before they could be collated, the charts could be drawn up only weeks or months after the day to which they referred. But if FitzRoy could find typical recurring weather patterns among those charts, it might be possible to infer from the barometric observations on land at least something of what was going on out in the ocean. It was FitzRoy who gave these weather maps the name 'synoptic' charts, since they provide a synopsis of what the weather is like at the same instant across a wide area of the globe.

Just when things seemed to be developing nicely, though, FitzRoy was distracted by the kind of political shenanigans that were so characteristic of Britain in the mid-nineteenth century.[9] In February 1855, the coalition government headed by the Earl of Aberdeen fell, and Viscount Palmerston became Prime Minister for the first time. Naturally enough, one of his supporters, Lord Stanley, became President of the Board of Trade, and, equally naturally for the times, he in turn appointed a protégé, one Lieutenant Simpkinson, to the vacant post of FitzRoy's deputy. FitzRoy found Simpkinson unsatisfactory, and demanded his dismissal. Preferring to jump before he was pushed, Simpkinson resigned; but Stanley refused to let FitzRoy have another clerk in his place. FitzRoy's response was to make do, as he had already been making do since August 1854, with just the three clerks, and to appoint the ablest of them, Pattrickson, as his deputy. But there was a snag. About the same time, as part of a larger reorganisation of the Board of Trade (and thanks to 'interest'), another of FitzRoy's clerks, Babington, was promoted from Supplementary Clerk to Junior Clerk, and therefore became senior, in the eyes of the Civil Service, to Pattrickson. FitzRoy worked around the problem by treating Pattrickson as his deputy and ensuring that he received a higher salary because of his skills as a draughtsman; Babington, a sensible young man, was quite happy to go along with this, but the third clerk, Townsend, refused. He had to go to another Department, and was briefly replaced by another junior, who also refused to accept the situation and also had to go. Eventually, FitzRoy got his way, with Pattrickson accepted as his de facto deputy. But it was at the cost of an immense waste of his time and effort, which would have been better spent, but for the petty-mindedness both of his superiors and some of his juniors, on developing the meteorological work. It is surely significant, though, that not only was FitzRoy willing to go to such lengths to look after his assistants, but that Babington and Pattrickson responded with equal loyalty to him and to each other. It was the same bond between FitzRoy and his subordinates, albeit on a smaller scale, that Darwin had noticed on the *Beagle*, and which echoed the loyalty shown by Graves and Skyring on the first *Beagle* voyage.

FitzRoy was, by the standards of his day, an enlightened 'employer' in many ways. In a memorandum dated 13 June 1859 he sets out the case for his assistants to have an extra half-day leave each fortnight (that is, alternate Saturday afternoons off) because he believes that as a result they 'will have better health, and do better work for their public pay'. To modern eyes, though, this liberality indicates just how hard both FitzRoy (who made no suggestion that he might himself take alternate Saturday afternoons off) and his subordinates worked. As if all this were not enough, alongside (and in addition to) his work at the Meteorological Office FitzRoy became a member of the Committee of Management of the Lifeboat Association, where he was active once again in doing his best to ensure the safety of sailors. As he put it to the Secretary of the Association:[10]

> Your work is of an affirmative character; there can be no misgiving about the work of the life-boat in saving a shipwrecked crew – it is palpable to everybody; but in respect to my work on the coast, it is somewhat of a negative character; I try, by my warnings of probable bad weather, to avoid the need of the life-boat.

The bureaucracy of the nineteenth-century British government system may have been tedious and frustrating at times, a far cry from the freedom FitzRoy had to make appointments as he chose when in command of his own ship on the other side of the world, but its wheels did inexorably grind out something that must have pleased him greatly around this time. In those days, the system of promotion from Captain to Admiral was still a matter of seniority. Once a naval officer was promoted to the rank of Captain (or 'made Post'), he knew that provided he lived long enough and enough of the existing Admirals and more senior Captains above him in the Navy lists died, he would eventually become an Admiral in his turn. There were obvious flaws with the system – a man might be rather old for the job by the time he became an Admiral, or simply not fit for such high rank – and by the 1850s there was a growing lobby (which included FitzRoy's old Lieutenant, Sulivan, now himself a senior officer) for reform. But the Navy had ways around the flaws – particularly talented Captains could be given the rank of Commodore (if necessary, temporarily) and put in charge of several ships, making them Admirals in all but name and entitled to give orders even to Captains with more seniority than themselves; and there was ample scope to ensure that the incompetent Admirals never went to sea in spite of their rank. For a Captain like FitzRoy, though, you could practically count the days towards the inevitable promotion by keeping track of the more senior officers falling off their perches. His promotion, to Rear-Admiral on the Reserved list, duly came in 1857, raising his status at least a notch or two.

This must have been small consolation, though, for two recent events that had hit FitzRoy hard. On 28 August 1856, his eldest daughter Emily, just eighteen years old, died. Such early deaths may have been more common in Victorian times than they are today, but the father's grief at the event, coming only four years after the death of the girl's mother, will have been no less heartfelt for that. FitzRoy's characteristic response was to fling himself into his work, staying even later at the office and keeping his mind occupied to blot out the grief. Remember the letter he wrote to Fanny describing his reaction when he learned of his own father's death:

> I did not shut myself up, nor neglect my duty for a minute, – I found that the more I employed myself and forced occupation, the easier I got through the day. – My worst time was when alone and unemployed . . . What a life this is – the pains are far greater than the pleasures – and yet people set such a value upon existence, as if they were always happy.

Those words apply equally well to the months following Emily's death. Less than two years later, FitzRoy's half-brother Charles died, in February 1858 at the age of sixty-one; Robert and Charles had never been close, but here was another reminder of human mortality. Not that FitzRoy took that as an indication that he ought to slow down – as we have mentioned, in 1859 he was elected as a member of the Committee of Management of the Royal National Lifeboat Association, a post in which he was as active as he was with all his commitments, which, as we shall see, began to mount up around this time.

In the same year that FitzRoy lost Emily, Admiral Beechey, the head of the Marine Department at the Board of Trade, and FitzRoy's immediate superior, died. On the positive side, Beechey and FitzRoy had never seen eye to eye on the responsibilities of the Meteorological Office, and this brought an end to some of the bureaucratic difficulties FitzRoy had encountered in his first years at the Board of Trade. Not unnaturally, he applied for the vacant post (its official title was Chief Naval Officer at the Board of Trade, or Professional Officer for short); but he was turned down, and in 1857 the post went to James Sulivan, FitzRoy's one-time protégé. This must have rankled, at least a little, but Sulivan had too much respect for his old Captain to allow the situation to be an embarrassment. A contemporary[11] tells us that:

> His relations with his old chief, Admiral FitzRoy, were interesting. When he came to the Board of Trade, FitzRoy was at the head of the Meteorological Office, and it was the duty of the officer in the place to which Sulivan was appointed to superintend FitzRoy's doings. But Sulivan made it a condition of his appointment that he should not be obliged to direct FitzRoy – a sagacious condition, considering the previous relations and the individual characters and tempers of these two distinguished men. Sulivan's respect and admiration for FitzRoy as an unrivalled sailor and a devoted public servant were unlimited.

The upshot was that FitzRoy received a much greater independence, reporting directly to the Secretary of the Board of Trade, with no Professional Officer breathing down his neck. Whatever FitzRoy may have felt at being passed over for this promotion, from our perspective it was undoubtedly good for the development of weather forecasting that he stayed where he was and put his greater freedom to good use. Just two years later came the event that would give him the leverage to bring his long-desired storm warning system into operation.

At the 1859 meeting of the British Association for the Advancement of Science (BA), the Council passed a resolution 'praying the Board of Trade to consider the possibility of watching the rise, force and direction of storms and the means for sending, in case of sudden danger, a series of storm warnings along the coast'. At that time, Prince Albert was President of the BA, and over the course of one of the hottest summers on record, he arranged for two meetings at Buckingham Palace to discuss the idea. But their deliberations were overtaken by events on the night of 25–26 October 1859, when a severe storm wrecked the modern iron ship *Royal Charter* on the coast of Anglesey Island, off northwest Wales, with very few survivors. FitzRoy's study of the circumstances surrounding the disaster convinced him that a warning could have been issued in time for the vessel to take shelter, and he made his case to the 1860 meeting of the BA, held in Oxford,[12] and in a paper published in the *Proceedings of the Royal Society* (volume 10, pages 222–4, 1860). The upshot was that on 6 June 1860 he was at last given formal permission to start a storm warning service to shipping, and by 1 September he had placed thirteen sets of instruments at locations around the British Isles.[13] Following a system already established in France by le Verrier, FitzRoy put the instruments in the charge of telegraph operators, so that they could make the observations and send them on to London routinely, without involving a third party and ensuring speed of communication. Everything worked from the word 'go' (a notable tribute to FitzRoy's powers of organisation and to the practicality and simplicity of both his barometer design and the instructions contained in his barometer manual). As a first step, the observations were made (free of charge) at 9 a.m. each day and immediately transmitted (at a special cheap rate) by the telegraph companies to the Meteorological Office in London. By swapping data with Paris, FitzRoy also gained further daily observations from six sites on the Continent.

The first phase of the operation, until February 1861, was devoted to keeping records and making sure that everything worked. As a result, FitzRoy was technically just beaten in the issuing of storm warnings by the Dutchman Christoph Buys Ballot, of Utrecht, who had founded the Netherlands Meteorological Institute in 1854; he gave warnings in 1860 for a small region of the Netherlands, using data from just four observing stations. But this was not a significant breakthrough, and Buys Ballot is best remembered in meteorological history for the law which bears his name, which tells you that if you stand with your back to the wind, in the northern hemisphere you will have high pressure on your right and low pressure (where the centre of a cyclonic storm and the strongest winds are likely to be found) on your left (the pattern is reversed in the southern hemisphere). The big step in storm forecasting came on 6 February 1861, when FitzRoy issued his first warnings, which were conveyed to shipping using a set of visual signals designed by him and hoisted at shore stations. The choice of warning signs itself indicates FitzRoy's thoroughness and attention to detail, and have never been bettered – they involve combinations of a cylindrical drum and a cone, hoisted singly or in pairs, with the great advantage that because of their symmetry the signs look the same from any direction. Although at first some vessels ignored the warnings and suffered the consequences, within weeks lives were being saved by the new system. France was quick to take it up, with other countries following suit.

This should have been not so much the pinnacle of FitzRoy's career as his arrival on a high plateau, with the value of his storm warning system widely recognised, and the opportunity to develop his meteorological ideas further for the public good, with his superiors at the Board of Trade and the Admiralty appreciating the value of this work and smoothing his path, even if he did not expect, nor seek, public acclaim. But just when FitzRoy should have been feeling secure professionally, he was thrown into an inner turmoil by the publication of the *Origin of Species* in November 1859 – just at the time when FitzRoy was investigating the *Royal Charter* storm and beginning to convince the authorities of the value of a system of telegraphed storm warnings. As if he didn't have enough on his plate at this time, in December 1859 FitzRoy entered into a correspondence in the columns of *The Times*, using the pen-name 'Senex',[14] in an attempt to refute claims by palaeontologists that flint axes found on the banks of the River Somme dated from 14,000 years ago – an age in conflict with the age of the Earth famously inferred by Archbishop James Ussher, from counting the genealogies in the Bible, as no more than 6,000 years. FitzRoy had long since been convinced of the literal truth of the Bible, to the extent of suggesting that giant animals such as the Mastodon had been unable to survive the Flood because they were too big to get into Noah's Ark; in January 1863, in a letter to John Herschel, he would write: 'I find Astronomy and Geology the *most* convincing proof of old Testament *inspiration* – (if *fairly* read – and thought of).' In spite

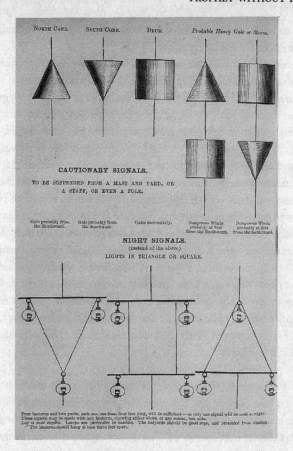

FitzRoy's warning signals

of these beliefs, and not least because of Darwin's reticence about going public with his ideas on evolution, the two former shipmates had kept in touch, with FitzRoy occasionally calling in on Darwin at his home in Kent when he happened to be passing that way, and Darwin writing a letter of sympathy to FitzRoy on the loss of his daughter in 1856. FitzRoy's last visit to Down House, Darwin's home, took place in the spring of 1857. But the publication of the *Origin* and the widespread acclaim it received destroyed what was left of their friendship, and caused FitzRoy 'the acutest pain'.[15]

In a letter to the editor of the *Athenaeum* (primarily written in connection with some correspondence about ships' magnetism) on 29 November 1859 FitzRoy refers to the review of the *Origin* the magazine had carried on 19 November, with what he regarded as a justified critique of 'my poor friend and five year messmate Charles Darwin'.[16] The tone of the remark is one more of sorrow than of anger, and this accurately reflects FitzRoy's feelings about the whole evolution business. He was genuinely sorry that Darwin had followed what he saw as a false path, and sure that his fate would be to burn in Hell. Any anger FitzRoy felt was directed at himself, not at Darwin – what made things worse was the knowledge that Darwin had made the observations on which his (to FitzRoy) blasphemous theory was based during the second voyage of the *Beagle*, where he had been present as FitzRoy's guest; so FitzRoy was to a large extent to blame, in his own eyes, for unleashing this evil upon the world. In this context, Darwin's explanation of the background to his theory, given in the very first words of the Introduction to the *Origin*, must have struck FitzRoy hard:

When on board HMS *Beagle*, as naturalist, I was much struck with certain facts in the distribution of the inhabitants of South America, and in the geological relations of the present to the past inhabitants of that continent. These facts seemed to me to throw some light on the origin of species – that mystery of mysteries, as it has been called by one of our greatest philosophers.

'When on board HMS *Beagle* . . .': there was the link with FitzRoy, plain for all to see. And there was another factor which gnawed away at him over the next few years – the beginning of the process whereby FitzRoy became known merely as Darwin's Captain, the sailor who happened to be in command (with the implication that he was only the driver, and any other sailor could have done the job) of the ship now famous for its connection with the naturalist, not for its real work of surveying. Although FitzRoy never sought public acclaim, life in the lengthening shadow of Darwin must have been particularly hard at a time when he might reasonably have expected his own achievements to be making something of a splash.

All of this was brought home to FitzRoy at the BA meeting in June 1860. Before then, he had been aware of the fuss aroused by the *Origin* (what Londoner of the time could not be?), but he had been preoccupied with his work in the wake of the *Royal Charter* storm, and had imagined that Darwin's ideas were widely regarded in the same way that he regarded them, as, to say the least, misguided. In Oxford, he found to his horror that he was wrong. His own talk, on 'British Storms', was presented to the meeting on Friday 29 June. This is a key paper in the history of meteorology, setting the *Royal Charter* storm in its historical context and, as we have seen, pointing the way to the first successful storm warning system in the world. But like everything else in Oxford that week, it was completely overshadowed by the debate about evolution. This centred on two papers, one read to the meeting on the Thursday, the day before FitzRoy's talk, and the other on Saturday 30 June. The first was on 'The Sexuality of Plants, with reference to the views of Mr Darwin'; the second on 'The Intellectual Development of Europe, with particular reference to Mr Darwin's work on the Origin of Species'. The first provoked passionate debate about Darwin's theory, and focused even more attention on the Saturday talk, attention fired by the news that Samuel Wilberforce, the Bishop of Oxford, would be attending, to put the Darwinists in their place (he was known behind his back as Soapy Sam, both from the slipperiness of his debating skills and his habit of rubbing his hands together while speaking, as if washing them). Darwin himself, something of a reclusive invalid by now, would not be attending; but he had an able champion in Thomas Henry Huxley, an impassioned supporter of the idea of natural selection, who became known as 'Darwin's Bulldog'.

The debate on that Saturday in Oxford has become the stuff of legend, and like all legends it is hard to pick out the nuggets of truth from the exaggeration and hyperbole. But it is generally accepted that Soapy Sam was savaged by Darwin's Bulldog. Following the presentation of the advertised talk, Wilberforce was called upon to address the packed audience, and exercised all his famous skills as an orator for half an hour, denouncing Darwinism and attacking Huxley and the other supporters of evolutionary ideas. It was a typically Soapy performance, with selective use of facts and the setting up of straw men to be knocked down. But at the end he made his big mistake. Turning to Huxley, he said:

> I should like to ask Professor Huxley, who is sitting by me, and is about to tear me to pieces when I have sat down, as to his belief in being descended from an ape. Is it on his grandfather's or his grandmother's side that the ape ancestry comes in?

Huxley rises to the occasion, after demolishing Wilberforce's points one by one, with one of the most famous responses in science:

> I should feel it no shame to have risen from such an origin. But I should feel it a shame to have sprung from one who prostituted the gifts of culture and eloquence to the service of prejudice and falsehood.

The rest of the debate had something of the feeling of 'after the Lord Mayor's show'. One of the few speakers from the body of the hall to oppose Huxley was FitzRoy. Standing (according to some accounts, with a Bible in his hand), he said that he 'regretted the publication of Mr. Darwin's book and denied Professor Huxley's statement that it was a logical arrangement of facts'. He said that he 'had often expostulated with his old comrade of the *Beagle* for entertaining views which were contradictory to the First Chapter of Genesis'.[17] But nobody seemed to care, and FitzRoy realised just how firmly established Darwinian theory already was. Horrified, he went home to London, and threw himself back into the meteorological work, no doubt gaining some consolation from the early success of the storm warning system. The spirit was undoubtedly still willing, but FitzRoy was just coming up to his fifty-fifth birthday, and had been working long hours since he was twelve; the physical and mental strain, exacerbated by the continuing attention for Darwin's ideas, soon began to take its toll.[18]

Over the next few months, as we have seen, the storm warning system was tested, and came into action in February 1861. A lesser man (or a less driven man), now into his late fifties, might have felt that he could take things easy for a while after that; but in addition to his official, and now routine, duties FitzRoy began working hard on two pet projects. The first was to develop the techniques pioneered for his storm warning system to provide the world's first regular system of daily weather forecasts – a term which he invented. Although this far exceeded the brief (such as it was) that he had been given when he took up his post, the logic was inescapable to FitzRoy, and makes sense today. His job (as he saw it) was to warn seafarers about hazardous conditions at sea, and the best way to provide warnings of storms was to look at the whole pattern of weather around the British Isles (for a start), develop an understanding of how it was changing from day to day, and use that as the basis for storm warnings. It all required a lot more data, from more observing stations, but FitzRoy had been steadily building up his network since 1857, when the Meteorological Office had started lending barometers and their instruction booklets to fishing villages and other coastal centres; this network had been further upgraded with the help of private benefactors, such as the Duke of Northumberland who paid for fourteen barometers to be sited along the northeast coast of England. And FitzRoy had been building up his staff, after overcoming the troubles associated with Babington's promotion. By 1862, he would have nine people working under him, although only the ever faithful Pattrickson and Babington were on the regular staff of the Board of Trade. With more people in the Office, and deliberately cutting down the collection of statistics, FitzRoy had the manpower he needed to move on to the forecasting work that he had long dreamed about.

FitzRoy has left us a detailed description of how the forecasts were prepared. Observers at each of his main stations recorded the temperature, pressure and humidity of the air, the wind speed both at ground level and (from studying cloud movements) at higher altitudes, the state of the sea, how all these parameters had changed since the previous observations, and the amount and kind of precipitation. The information was sent from each observing station at 8 a.m.,[19] and:

At ten o'clock in the morning, telegrams are received in Parliament Street, where they are immediately read and reduced, or corrected, for scale-errors, elevation, and temperature; then written into prepared forms, and copied several times. The first copy is passed to the Chief of the department, or his Assistant, with all the telegrams, to be studied for the day's *forecasts*, which are then carefully written on the first paper, and copied *quickly* for distribution.

At eleven – reports are sent out to the Times (for a *second* edition), to Lloyd's, and the Shipping Gazette; to the Board of Trade, Admiralty, Horse Guards, and Humane Society. Soon afterward similar reports are sent to other afternoon papers: and, *late in the day*, copies, more or less *modified* in consequence of telegrams received in the afternoon, are sent out for the [first editions of] next morning's papers.

In other words, the first forecasts left the Meteorological Office just one hour after the raw data came into the building; this would be an impressive achievement even today, but FitzRoy and his assistants didn't even have typewriters to speed their work, let alone photocopiers or word-processing computers. Everything had to be written by hand, and the only computers available were human beings. FitzRoy was at pains to point out that all this came at very little cost to the taxpayer, 'the observations and their telegraphic communication to London having been authorised and paid for by Parliamentary vote, chiefly for scientific purposes, out of which these *additional* practical measures have legitimately grown – not at *great* additional charge'.

His attention to detail is highlighted by one feature of these first forecasts which looks odd to modern eyes, but which had a perfectly natural explanation given the pace of life in Victorian Britain. From the outset, FitzRoy issued forecasts in pairs, not just for the following day (a difficult enough prognostication, you might think, given the state of the art) but for the next day as well. At first sight, this looks like nothing more than bravado; but FitzRoy explains:

While newspapers are being printed and are travelling to distant towns, time escapes so fast, that the day and its forecast too often arrive at a distant place together, and then of course there is no value in such information – the day with its *weather* being *present*. For this there is yet no remedy; but in some measure the *second* day's forecast may make up for the retardation of the first, if fairly noticed and compared.

There lay the rub. All too often, once the novelty of the forecasts had worn off, what FitzRoy was trying to do was not 'fairly noticed and compared' by his critics, who poured scorn on the forecasts that failed, ignored the ones that were correct, and (to his particular chagrin) failed to see the link between the proven success of the storm warning system and the underlying understanding of weather patterns that he had developed. The value of the warnings, he emphasised, 'has been fully proved and acknowledged; but the forecasts, which are their actual foundation, are not yet so generally noticed, not being duly compared, or fully appreciated'. But he was to have one more moment in the sun before those criticisms began to wear him down.

FitzRoy's second big project in the early 1860s, which had to be completed in his own time, was a weighty volume, addressed to the general reader, summarising the state of understanding of meteorology at the time and putting forward FitzRoy's own ideas about weather forecasting. When he sent a copy to Herschel in March 1863, the covering letter explained the circumstances in which it had been written: 'When you know that it was begun at Brighton, in my so-called holiday, on the 10th of last August, and was circulated to the public in December – your remark will probably be – "Better to have taken more time".' But FitzRoy had no more time – unlike Darwin, he was not a gentleman of private means who could give all his attention to such a project, but a paid public servant, with a powerful sense of duty that prevented him working on his own book during office hours. Of course, much of the material had been gathered before he began writing; but even so this indicates just how little rest (let alone holidays) FitzRoy took in the 1860s. The *Good Words* obituary gives a little more detail:

It was a rare thing for him to take a holiday, and even on this occasion it did not imply total absence from his duties. He was within easy reach of the office, going there and returning frequently the same day. This work [on the book] was continued, and completed at his own residence on his return home, evening after evening, through successive nights. His overstrained mind never entirely recovered [from] this pressure. The late work, destroying his night's rest, soon told its tale; and from that time he was totally unable to write in the evening, or even to read for a few minutes without falling asleep. In vain he struggled against this propensity, trying every possible means to overcome it, but without avail. A more serious inconvenience resulted also from this overstrained pressure on the brain – an increasing deafness, from which he had slightly suffered for many years past, especially when much fatigued. He consulted aurists, but derived no benefit, and he began to dread lest he should become stone-deaf.

The epic volume resulting from this supreme effort was plainly titled *The Weather Book*, and as FitzRoy wrote at the beginning of the book, 'under so plain a title neither abstruse problems nor intricate difficulties should be found. This popular work is intended for many, rather than for few, with an earnest hope of its utility in daily life.' He also gives a clear indication of the way his own views on what he should be doing at the Board of Trade differed from those of the people who had appointed him. 'Until lately,' he writes, 'meteorology had been too statical in practice to afford much benefit of an immediate and general kind. Indefinitely multiplied records only tended to make the work of their utilisation discouraging, if not almost impossible.' And this from a man whose official title was still Meteorological Statist!

The book certainly reached its target audience, and the first edition of 1862 was followed by a second edition the following year; FitzRoy was paid an outright fee of £200 for the copyright, and the book was priced at fifteen shillings. Even though he was now a successful author in his own right, however, the sales in no way matched the runaway success of the *Origin* and other books by Darwin, which must have stirred memories of the irritation FitzRoy had felt when Darwin's contribution to the *Narrative* of the *Beagle* voyages had been split off and published separately to wide acclaim.

Inevitably, some of the ideas in *The Weather Book* have turned out to be wrong. FitzRoy's own pet hypothesis of the causes of weather variations, discussed at the end of the book, certainly does not stand up to inspection today (so there is no point in discussing its details), and came in for severe criticism at the time, notably from Herschel. FitzRoy thanked Herschel for his private comments, which discouraged him from offering the ideas in a paper to the Royal Society, for fear of making a fool of himself. This must have hit him hard, and added to his growing sense of failure. But today, reading *The Weather Book* almost 150 years after it was published, it is striking how much is right, and how well FitzRoy understood the broad picture. Examination of hundreds of synoptic charts, for example, had convinced him of the reality of troughs and crests (what are now called ridges) of pressure associated with different kinds of weather, and provided '*real* proof of *areas* of depression', our lows, as well as showing him how the overall pattern of weather moved from west to east. All these factors went into his daily weather forecasts. Although these were presented in the form of words, not charts, FitzRoy would have had no difficulty 'reading' a modern synoptic chart. He describes his work at the Board of Trade up to the issuing of the first published weather forecasts, and offers convincing proof that his system of forecasting and storm warnings works, quoting from a meeting of the shareholders of the Great Western Docks at Stonegate, Plymouth, where in 1862 'it was stated officially that "the deficiency (in revenue) was to be attributed chiefly to the absence of vessels requiring the use of the graving docks for the purpose of repairing the damages occasioned by storms and casualties at sea"' since mariners had started to take note of FitzRoy's warnings. And among the many diagrams in the book there is a beautiful representation of a typical weather pattern over Britain and nearby Europe, based on FitzRoy's synoptic charts but looking for all the world as if it were based on satellite photographs of real weather systems.

Everything is rounded off with some dramatic examples of the power of storms at sea, some from FitzRoy's personal experience, others from reports by other sailors. One of the last of these anecdotes serves as a warning to mariners who fail to heed the danger signs. It describes how, on the way home from New Zealand with his family on board the

David Malcolm, FitzRoy was in the hands of a Captain who had the most casual attitude to navigation, and didn't even carry a single barometer. Anchored at the western end of the Magellan Strait on 11 April 1846, this man was preparing to go to sleep, happy that his ship was safe riding to its lightest anchor on a single cable, when FitzRoy made so much fuss about the fact that his two private barometers were falling sharply that the lighter spars were taken down from aloft, and a heavier anchor veered out on a chain. With the Captain safely asleep below, as the barometers fell to just above 28 inches, at midnight FitzRoy persuaded 'a good officer and a few willing men' to let go a second anchor as well, even though it was a beautiful, still and clear moonlit night. The storm hit at 2 a.m., and even with both anchors the ship still dragged across the harbour, coming 'within a stone's throw of sharp granite rocks astern, at some distance from land, near the most exposed outer point of the harbour', before the wind abated. Without the precautions FitzRoy had insisted upon – without his barometers and his understanding of what they foretold – the ship would have been lost with all hands.

With the publication of *The Weather Book* and the interest roused by the publication of the first regular series of weather forecasts in the newspapers, 1862 marked the pinnacle of FitzRoy's career, topped off by his promotion to Vice Admiral in 1863. He must also have been pleased with the progress his son was making in the Navy, where he had been serving as a Lieutenant in the Far East. In a letter to John Herschel dated 21 April 1862, among many scientific topics he discusses FitzRoy mentions that his son has visited the Great Wall of China (and sent some bricks from it back to England!) and comments that in Japan 'compasses [have] needles *marked* on the *North* end – unlike the Chinese, *marked* on the *South*'. Clearly, FitzRoy maintained a lively interest in a wide variety of scientific topics beyond the requirements of his work at the Meteorological Office. In 1864, he was elected a Corresponding Member of the French Academy of Sciences in recognition of his work, an honour that he was particularly proud of.

But any feeling of triumph was short-lived, and in his lifetime FitzRoy received less recognition in his own country. He certainly had strong support for his work from two important quarters – the storm warnings in particular were widely appreciated by mariners, and had even proved their commercial value (to shipowners, if not to dockyards!), while the Royal Society continued to support his scientific efforts, speaking up on FitzRoy's behalf in 1863, when the Board of Trade expressed concerns about the increasing costs of his Office.[20] Even *The Times*, which frequently commented on the forecasts with heavy-handed humour, agreed that these first steps towards accurate forecasting were worthwhile. And he had patronage from the highest in the land – Queen Victoria got into the habit of consulting him before deciding when to set out on the short crossing of the Solent to reach her house on the Isle of Wight, and on at least one occasion a forecast was requested by the Palace when one of the princesses was planning to cross the Channel, from Folkestone to Boulogne. A copy of FitzRoy's forecast sent back to the Palace for the day in question, Friday 4 March 1863, was preserved by his widow: 'Weather on Friday favourable for crossing – Moderate – mild – cloudy, fine, perhaps showery at times.'[21]

In a Foreword to the book by Mellersh, Nora Barlow, Darwin's granddaughter, describes a meeting with Laura FitzRoy, Robert's daughter by his second marriage, that took place in 1934. In that interview, Laura describes how when she was a young girl living with her parents in Kensington, she heard the door bell ring one day, and ran downstairs to open the door, expecting to see her parents returning from Communion. To her consternation, it was actually a Queen's Messenger, wanting a weather forecast for the next day for Victoria, who was planning to cross the Solent to the Isle of Wight:

> The charm of Miss FitzRoy's personality and her appearance in her white cap and her shawl as she told me this story, still remain vividly stamped on my mind. After my visit I felt that I could realize more fully how the intransigence of the *Beagle*'s Captain could have been combined with an almost tortured sense of right and wrong; how righteous indignation and desire to punish was mitigated by his sense of generosity; and how compassion was near the surface, except perhaps towards himself.

We don't quite know where Nora Barlow got the idea that FitzRoy ever had 'a desire to punish' anybody; but let that pass.

In spite of all this support and the royal seal of approval, there were three kinds of critics whose attacks now loomed large in FitzRoy's mind. First, there were those who criticised him on scientific grounds. This was fair enough; rival ideas about how the weather worked existed, and the correct scientific procedure was to debate the issue, make observations, and see which model turned out to be right (even if the debate got a little heated along the way). Unfortunately, though, along with people such as the lunarists, who claimed to have found a link between the moon and the weather, this group included Francis Galton, who happened to be a cousin of Charles Darwin and was also General Secretary of the BA, and pioneered the use of statistical methods in science. Galton had a serious interest in meteorology – he gave the name 'anticyclone' to high pressure systems, and was responsible for the publication of the first weather map in a daily newspaper, in *The Times* in 1875. Quite apart from the Darwin connection (Galton was an enthusiastic supporter of his cousin's ideas on evolution), there had been antipathy between Galton and FitzRoy since 1858, when Galton asked FitzRoy to offer to the Navy his invention of a 'hand heliostat', a signalling device based on mirrors, and FitzRoy had passed it to the Hydrographer's Department with the comment that it was not thought to be very practical.[22] But Galton was now in an influential position in British science, and would soon use that influence to the detriment of FitzRoy's posthumous reputation.

The second group of critics was small but vociferous, and dominated by James Glaisher, who you might have expected to be all in favour of FitzRoy's work. The problem was, these people objected to the free provision of weather forecasts by the Meteorological Office, because they realised that if the forecasts were reliable then money could be made by selling them. In 1863, Glaisher was involved in an attempt to form a company called the Daily Weather Map Co. Limited, with the idea of selling a newspaper devoted to the publication of weather maps and other meteorological information. But the project came to nothing, not least because it would make little sense for people and businesses to subscribe to such a paper if they could get the same information from

The Times. The final group of critics included some shipowners who were sufficiently short-sighted to object to the loss of revenue that resulted when their ships stayed in harbour as a result of storm warnings, without making allowance for the money saved by not having the ships sunk.

There were other thorns in his side. Although the Meteorological Office itself checked the accuracy of its warnings, the Wreck Department of the Board of Trade carried out independent checks, the results of which were published as a Parliamentary Paper in 1864, much to FitzRoy's annoyance and prompting him to write to *The Times* (9 June 1864) pointing out that the report had not come from his Office. On the administrative side, there were also major problems with the way the money spent by the Meteorological Office was being accounted for. As always, FitzRoy's attitude was that you had to spend what was necessary when it was necessary to get the job done, and keeping track of everything on paper was never his forte. FitzRoy's attitude that the public good came first and you worried about the money afterwards was exactly the opposite of that of the Board of Trade, and although nobody doubted his integrity or suggested that there had been any fiddling of the accounts, this became a major bone of contention in the last years of his life.[23] Among the FitzRoy papers at the Public Record Office there is the draft of a letter to H. R. Williams, the accountant with the Board of Trade, dated 20 March 1863 and showing all too clearly the way things were getting on top of FitzRoy by then. His handwriting is large and hurried, but starts legibly enough before becoming a mass of crossings-out and corrections as his mind is clearly racing on much faster than his hand can write.[24] We do not have the letter from Williams to which he is replying, but clearly the accountant has taken the time to draw FitzRoy's attention, as gently as possible, to some financial or bookkeeping irregularities. 'I am much obliged to you,' FitzRoy begins, 'for writing a private note instead of an official Minute about these *necessary* however annoying affairs.' But after mentioning some specific points, he continues:

> I am now, as it were, on sufferance . . . I have, for some time, been considering my own prudent course – as health will not withstand

continued delays, and mortification. But these feelings are behind the mask – at this moment – I only express them, privately, to yourself.

But then comes the conclusion of a man at the end of his tether, as he says that he feels 'Like an Israelite disgusted by brick making *without straw*' before reining himself in to end with the civil salutation, 'though still cordially yours Robert FitzRoy'. There is no evidence that the letter was ever sent, but nothing could more clearly show FitzRoy's state of mind as he saw himself attacked from all sides.

He was even attacked for things he had no control over. There was the problem that any storm warnings issued, including those from abroad, were being attributed in the public mind to FitzRoy's Office, and although these were more likely to be misguided than malicious, when (as was often the case) they turned out to be false his reputation suffered. But the most disturbing criticism came from Maury, who, as a supporter of the Confederacy in the American Civil War, had been forced to flee his home and was now based in Europe. In two articles published in Paris, Maury attacked the basis of FitzRoy's forecasting system. This wounded FitzRoy deeply, spurring him to a furious spate of writing letters and articles in response. One unfortunate consequence of this was that criticism made by FitzRoy of the Director of the Imperial Observatory in Paris, in a letter he intended as a private communication to the head of the French Meteorological Service, Marié Davey, was published, leading to a cooling of relations between FitzRoy and his French counterparts. But in spite of their professional differences, FitzRoy remained concerned about the fate of Maury's family, embroiled in the Civil War and thrown out of their home.

By now, it seems, he worried about everything. He was in financial difficulty (only partly eased in 1863 when his salary was increased from £600 to £800, the same as Sulivan's), largely because of the amount he had expended from his private funds for the public good over the years; his health was failing, in no small measure because of the long hours he had always put in at work; and he was going deaf. Behind all this were his concerns about his own role in giving Darwin the opportunity to come up with the idea of evolution by natural selection, and, no doubt, his disappointments concerning the Fuegians, so lately in the news again for all the wrong reasons (see Appendix I).

After about 1862, FitzRoy aged rapidly and visibly; by the end of 1864, when he was well into his sixtieth year, the work of the Meteorological Office, including the forecasting, was largely being carried out by Babington, who had taken over as FitzRoy's deputy when Pattrickson moved on in 1863. And even *The Times* seemed to be having second thoughts about the value of his forecasts. On 18 June 1864, the Thunderer commented that 'Whatever may be the progress of the sciences, never will observers who are trustworthy and careful of their reputations venture to foretell the state of the weather.' The article then gets personal, attacking FitzRoy's use of jargon and continuing: 'What he professes, so far as we can divine the sense of his mysterious utterances, is to ascertain what is going on in the air some hundreds of miles from London by a diagram of the currents circulating in the metropolis.' Just so – but the fact that this was seen as laughable in 1864 shows just how far ahead of his time FitzRoy was. No wonder the strain had begun to tell. By the beginning of 1865, his wife, Maria, and his doctor were sufficiently worried about his mental and physical well-being that the family moved out of central London, intending to stay for a few weeks in the quiet suburb of Norwood. In April, Maria wrote to a relative that:

> The Doctors unite in prescribing total rest, and entire absence from his office for a time. Leave has been given him, but his active mind and over-sensitive conscience prevent him from profiting by this leave, as he does not like to be putting the work he is paid to do upon others, and it keeps him in a continual fidget to be at his post, and the moment he feels at all better he hastens back, only to find himself unable to work satisfactorily when he gets there.[25]

The *Good Words* obituary reinforces this impression of a restless workaholic:

When advised to rest by those whom he consulted, he asked how much longer he could work without risk. On being told that immediate rest was necessary, he replied that when such and such letters were answered then he would rest. That time never came.

He had often said that 'he would rather wear out than rust out'. To die unflinchingly working to the last with his hand on the helm, was his wish; and this was realized to the utmost.

It's almost as if he was still trying to please his long-dead father.

So, even when officially on sick leave, he still insisted, when he felt well enough, on going in to the Office and making other visits to the centre of town. His last official action, as it transpired, was to send a weather forecast to the Queen for a planned crossing to the Isle of Wight.

News of the assassination of Abraham Lincoln, on 14 April 1865 (five days after the surrender of General Robert E. Lee) reached England in the last week of April, and seems to have triggered another bout of anxiety, both on behalf of Maury and his family and a general concern about the state of the world.[26] A last meeting with Maury, at the end of April, upset FitzRoy considerably, but unfortunately there is no record of what passed between the two men on that occasion. His last days were described by Mrs FitzRoy in a document still in the possession of the family:

Upper Norwood Friday, April 21st, in bed all day very ill, Saturday April 22nd in bed better. *Most grateful* for recovery. Sunday – up, weak and ill, unable to go out, but came downstairs, asked for his bible and prayer book, while we went to church; told me on my return that he had read as much as his mind was able to take in.

On Monday 24th, still very weak, able to take a short drive with Ad. Cary which did him good, and then sat out with me in the garden while the girls played at croquet.

Tuesday 25th. Much better. Would go to London – did no business at his office, came back in the afternoon early, met him accidentally near Mrs Thelmson's where we were going to pay a visit, would not come with us, but went home to get a cup of beef tea which I had ordered to be ready for him. Went to see A^d. Cary (not at home), waited for us, joined us and then went home. Evening played at whist which he seemed to like.

Wednesday the 26th, he went to London for a short time, not to his office.

Thursday 27th, a hot day, he started directly after breakfast for London in order to be back in time to meet General and Mrs Wood, came back just after 12 very tired out, and lay down to rest, came down to luncheon. Joined us and Colonel Smyth in a walk in the afternoon, had a good deal of conversation with Col. Smyth and myself about our affairs. He met Dr Hetty [Heatley] afterwards and had a conversation with him about his health, the last time Dr Hetty [Heatley] had any conversation with him. In the evening he seemed quiet and happy, talking tranquilly with me alone, and seemed to have made up his mind to stay here quietly and really take care of himself. Just before going to bed he received a letter from Mr Tremlett inviting him and myself to come and stay with him from Saturday till Monday to see the last of Capt. Maury. This note seemed completely to upset him, between desire to comply with his request and his just expressed wish of remaining quiet. Of course he did not sleep well that night; the only advice I gave him was to do that which would give his mind the greatest ease. On Friday morning he went to London to his office, came back again relieved at having written and refused the invitation, so he told me. And after luncheon he went to his room to write and called me urgently to come to him; when I came, I found him extremely distressed at the quantity of unanswered notes and invitations to public dinner which ought to have been answered long ago. I comforted him and helped him to answer two or three most pressing, and then he consented to go out with us for a little while – went to see Admiral Cary, who came back with him to the door; in the evening he again played at whist, which he seemed to like. Saturday morning after breakfast he came to me saying he had got a strong desire to see Maury again; I told him he had better gratify it if he had; he said he was totally incapable of exertion, and could only lie down and rest and asked me to make him comfortable, which I did. After luncheon he felt somewhat better, and set out to take a walk with the two eldest girls[27] while I went for a drive, in which he declined accompanying me, because he thought the walk would do him more good. When I came home I found that he had left them, and gone to London, and did not return till nearly 8 o'clock, worn out by fatigue and excitement and in a worse state of nervous restlessness than I had seen him since we left London. He seemed totally unable to collect his ideas or thoughts, or give any coherent answer, or make any coherent remark. After dinner he recovered a little and mentioned a circumstance which had impressed him awfully, which had occurred at Mr Tremlett's and also mentioned Mr Tremlett's having asked him to come again on Sunday. I expressed my surprise, as I had myself written to Mr Tremlett on Friday morning by the day post thanking him for the previous invitation, but telling him that the kindest thing his friend could do would be to leave him quiet, without tempting him to go up to London, as he never went there without being the worse, and never remained quiet here for a day or two without being decidedly better.

I offered to have another game of whist in the evening; he said there was nothing he should like better, but he had had such exciting conversations with Captn. Maury and Mr Tremlett that he could not divert his mind to any other subject. He generally went to sleep after dinner for a little time, but this evening he did not even close his eyes for an instant. When the girls had gone to bed he said to me he wished to talk over with me about his idea of going to London on Sunday to see Maury once again. I asked him if he had not wished him good-bye: he said he had. I then said I was very tired and sleepy, and the best thing we could both do was to go to bed, and talk over that the following morning. He agreed with me, saying how worn and tired I looked. I went away and then went downstairs again as he had not come up to his dressing-room. He thanked me for coming to see after him, and said he was coming up directly; he was standing up by the table, with the newspaper open before him, and was not long in coming to his room. I was in bed when he came to bed; he came round to the side where I was, asked me if I was comfortable, kissed me, wished me good night, and then got into bed. It was just 12 o'clock. I was soon asleep, and when I woke in the morning I said I hoped he had slept better, as he had been so very quiet. He said he had slept he believed, but not refreshingly; he complained of the light, and I said we must contrive something to keep it out. Just then it struck six. From 6 to 7 neither of us spoke, being both half asleep I believe. Soon after the clock struck 7 he asked if the maid was not late in calling us. I said it was Sunday, and she generally was later, as there was no hurry for breakfast on account of the train at 10 o'clock as there was on other days. The maid called us at ½ past 7. He got out of bed before I did, I can't tell exactly what time, but it must have been about ¼ to eight. He got up before I did and went to his dressing room kissing Laura as he passed through her little room, and did not lock the door of his dressing-room at first.

Then, on Sunday 30 April 1865, Robert FitzRoy bolted the door to his dressing room, picked up his razor and cut his own throat.

CHAPTER TEN

Aftermath

FITZROY'S OBITUARY IN the journal of the Lifeboat Association[1] refers to him as 'the skilful sailor, the travelled naturalist, the earnest Christian, and the best friend of the population which fringes our sea-girt isle'. Of all the testimonials published after his death, that is probably the one that he would have most appreciated. But those with less direct knowledge of how his work had helped sailors were less appreciative. His death brought to a head a crisis that would probably soon have confronted the Meteorological Office in any case. The bean counters of the Board of Trade were not convinced that the forecasts were providing value for money, and there was no doubt that FitzRoy had exceeded his brief in deciding to initiate a weather forecasting service. While Babington took over as acting head of the Office and a rather leisurely process of finding a replacement for FitzRoy went on, the Board, with advice from the Royal Society, set up a small committee of enquiry to look at the work of the Office. Since the chairman of that committee was Francis Galton, its conclusions cannot have come as a surprise to anybody. The report of the committee was presented to Parliament on 13 April 1866, and found fault with almost everything that FitzRoy had done, from his methodology to the way he presented his results. The report admitted that the storm warnings were 'of some use', and said that they were 'too important, too popular, and too full of promise of practical utility to be allowed to die', with the implication that they were not yet of practical utility, but found that the daily weather forecasts were of no value. As for the compilation of statistics from ships at sea that had been the raison d'être of the Office, Galton felt that FitzRoy had stopped collecting data much too soon, with at least three times as many observations required before the database would be of much use.

Galton knew enough about statistics to present numbers that backed up his case that even the storm warnings were not yet of much 'practical utility', however popular they might be, and these were accepted without question from such an eminent authority. But a close look at those numbers (analysed in the 1980s by Jim Burton) shows that this was essentially a smoke and mirrors job. Along a stretch of the northeast coast of England from Scarborough to Berwick, for example, there were nine observing stations maintained by the Department of Wrecks, whose statistics (rather than the Meteorological Office's own statistics) the Galton committee chose to use. When FitzRoy issued a gale warning for this stretch of coast, it went down in the records as nine separate gale warnings, and if just one of the observing stations was hit by a gale (Force 8 or above on the Beaufort Scale), that was recorded as one success and eight failures in the Galton analysis. This would be true even if all eight of the other observers recorded winds of Force 7 in the seventy-two hours following the gale warning (the period such a warning was in force for). In addition, the observers were based on land, where the wind was likely to be less severe than at sea, which was where the warnings applied. The modern practice makes such measurements over different regions of the sea, called sea areas. Counting by sea areas also amalgamates some of the Department of Wrecks observing sites, reducing the number of individual warnings to be counted. Burton found that using data from three days chosen at random in 1863, the Galton system gave 84 warnings correct out of 193 issued (a 44 per cent success rate), while the modern method gives 28 correct out of 37 (a 76 per cent success rate). Taking all the warnings issued for the month of December 1863, Galton has 236 correct warnings out of 387 (61 per cent success) while the modern method has 68 correct out of 76 (90 per cent). There are other, more subtle, flaws in the statistical presentation of the data in the Galton report, and it is hard to believe that Galton, of all people, was not aware of the way he was presenting the information to show FitzRoy in the worst possible light. Burton concludes that 'the report's figures were simply not fair by even the crudest of assessments, but they were official and were quoted widely, the general view of FitzRoy's work suffering in consequence'.

In the light of the Galton report, the Royal Society advised the Board of Trade to stop issuing the forecasts, and the last pair were published in *The Times* on 28 May 1866. Even Galton had not gone quite so far as to recommend withdrawing the storm warning service, but it was now decided that the operation of the Meteorological Office should be under the control of a committee appointed by the Royal Society, and the Council of that august body decided on the complete withdrawal of the storm warning service as well, on the grounds that it was based on largely empirical rules, and was therefore not suitable for administration by a scientific body. The term 'pompous asses' springs to mind, but the storm warnings were indeed stopped, on 7 December 1866, the day that Babington left his temporary command. But although it is always possible to lie with statistics, and empirical rules may be an anathema to a certain kind of ivory-tower scientist, the people whose lives depended on the storm warnings knew their value. The obituary in *The Life-Boat* had already commented that the warnings 'are looked upon as essential as a life-boat or a light-house at our ports; and the man who would neglect them is regarded as foolhardy by his brother seamen'. Now, the Board of Trade and the Royal Society were left in no doubt what the mariners themselves and harbour authorities felt about the cessation of the warning system, and the archives include among many similar communications a letter from a Fellow of the

Royal Society based in Cumberland, who writes, 'I trust the suspension will not be of long continuance, for the "warnings" are invaluable on this coast . . . Thirty-three warning telegrams have been received at this station during the current year, and in twenty-six instances a gale has followed from the Quarter indicated.'

The harbour authorities at Deal commented, 'There is but one opinion concerning the value of these signals. They have been the means of saving life and property to an immense amount.' And at the 1867 meeting of the BA, after a series of impassioned speeches a resolution urging the immediate restoration of the warnings was passed unanimously. In the face of such widespread criticism, the warnings were resumed at the end of 1867. Routine weather forecasts only began to be issued again ten years later, and it is probably true that in trying to provide such general forecasts, as opposed to specific storm warnings, FitzRoy was trying to cross a bridge too far. Even so, his precedent did help to ensure that when forecasts were resumed it was not on a commercial basis, but in line with his principles, as a free service provided for the good of all people. This aspect of FitzRoy's influence on the way the meteorological services of Great Britain and the world developed is seldom given the credit it deserves.

FitzRoy brought to the Meteorological Office the same qualities and character flaws that he had brought to all his work – a vision for the big picture, which swept him and his colleagues up and enabled them to get things done that might have seemed impossible, but an unwillingness, or inability, to bother with the nitpicking details of daily routine that left him open to attack by men more at home in the bureaucracies he served. He was moody and stubborn, but inspired great loyalty and affection from his subordinates and colleagues, even when there were disagreements between him and them. Even allowing for the tendency of a nineteenth-century obituary to praise Caesar rather than to bury him, the comments made about FitzRoy soon after his death reveal a man of principle, dedicated to public service, and give further clues to his ultimately suicidal character:

Never was there a man more free from personal vanity or self-love. He was always more ready to blame himself, than to censure others. Strictly conscientious in principle, nothing would make him swerve from what he considered the path of duty. The public claims were his first consideration; he regarded the time during office-hours as that of the Government by whom he was employed, and not at his own disposal. So scrupulously did he carry out this principle, that, during that time, he carefully avoided all personal matters, writing his own private letters after five o'clock, before his return home.

He was the most devoted of husbands, the tenderest of fathers, and a very warm and true friend.[2]

Not that FitzRoy was a clock-watcher in the other direction; he wouldn't hesitate to spend longer at the office if the need was there. The *Good Words* obituary adds that 'it is but the plain truth to say that he would exert himself in a friend's cause far more eagerly and indefatigably than in his own . . . if he had but one shilling left in the world, he would have given it away if touched by the cry of distress', and quotes the words of the then Admiral John Lort Stokes, describing FitzRoy as one 'in whom with rare and enviable prodigality, are mingled the daring of the seaman, the accomplishments of the student, and the graces of the Christian'.

The obituary by Sir Robert Murchison bluntly points the finger of blame at those responsible for FitzRoy's death:

In deploring the loss of this eminent man who was as truly esteemed by his former chief, the Prince of Naval Surveyors, Sir Francis Beaufort, as by his successors, I may be allowed to suggest that if FitzRoy had not had thrown upon him the heavy and irritating responsibility of never being found at fault in any of his numerous forecasts of storms in our very changeful climate, his valuable life might have been preserved.

And FitzRoy's earlier achievements were not forgotten. The Naval Hydrographer, Beaufort's successor George Richards, wrote to his widow that:

There is and ever has been one opinion among the officers past and present of this department as to the services of the late Admiral FitzRoy. No naval officer ever did more for the practical benefit of navigation and commerce than he did, and did it too with a means and at an expense to the country which would now be deemed totally inadequate . . . In a little vessel of scarcely over 200 tons, assisted by able and zealous officers under his command, many of whom were modelled under his hand and most of whom have since risen to eminence, he explored and surveyed the continent of South America . . .

The Strait of Magellan, until then almost a sealed book, has since, mainly through his exertions, become a great highway for the commerce of the world – the path of countless ships of all nations; and the practical result to navigation of these severe and trying labours, which told deeply on the mental as well as the physical constitution of more than one engaged, is shown in the publication to the world of nearly a hundred charts bearing the names of FitzRoy and his officers, as well as the most admirably compiled directions for the guidance of the seamen which perhaps was ever written, and which has passed through five editions . . .

His works are his best as they will be his most enduring monument, for they will be handed down to generations yet unborn.[3]

And let us not forget that, for all their differences in the later years of FitzRoy's life, Darwin's assessment in his own *Autobiography* concluded that: 'Fitzroy's character was a very singular one, with many noble features: he was devoted to his duty, generous to a fault, bold, determined, indomitably energetic, and an ardent friend to all under his sway.'

FitzRoy died in debt, a result of his generosity to those in need and his willingness, going right back to his time in command of the *Beagle*, to spend his private fortune for the public good. Probably the most accurate assessment of the extent to which he had spent his private fortune on the *Beagle*'s second voyage alone (not counting his costs in

New Zealand, for example) comes from the anonymous testimonial in *The Life-Boat* published on 2 October 1865, by someone who clearly had inside information and must have been one of the officers who sailed with FitzRoy. He points out that since 1825, there had been three expeditions to survey the coast of South America, each lasting five years:

> The second of these, under the command of Admiral FitzRoy, was [the] most successful and important. It opened up the route now usually taken through the Strait of Magellan, by which the dangerous navigation of Cape Horn is avoided, and it led to observations and discoveries which have made it an epoch in the recent history of science. Those well qualified to pronounce an opinion, say it may be compared to its advantage with any five years' survey in the records of the Admiralty, yet the expense to which the country was put by it was small. The first expedition to which we have referred cost 100,000*L.*, the last expedition 75,000*L.*, and that of Admiral FitzRoy only 40,000*L.* This saving to Government was, however, effected in a great measure at his own cost.

After detailing some of the reasons for these costs, notably the hire and purchase of auxiliary vessels (but not including the chronometers, and not mentioning that in addition to the survey FitzRoy had put his longitudinal chain around the Earth), the writer concludes that:

> His whole expenditure then, beyond the sum allowed at the Admiralty, amounted to 6,100*L.*, for which he was never compensated in any way; in fact he had to borrow money to discharge the liabilities he incurred in the public service.

This sum of £6,100 would be equivalent to at least £400,000 in modern terms. After FitzRoy's death, the government was belatedly shamed (not least by the efforts of Sulivan, who, according to Lord Farrer,[4] 'moved heaven and earth to get his services acknowledged') into making good some of the expenditure he had incurred in their service. On

20 June 1865, the Duke of Somerset (who was then First Lord of the Admiralty) wrote to Thomas Gibson, the President of the Board of Trade:

> I sent to Gladstone[5] the report of our Hydrographer on the late Captn FitzRoy's services. I could not add much to that report.
>
> There is some difficulty in deciding how to proceed.
>
> There are two families, and one, I am told, has got some small provision, while the other has nothing.[6]
>
> Then it is stated that Captn FitzRoy expended much of his own fortune in carrying on surveys &c. I think this should be inquired into so that a report could be made to the Treasury, otherwise similar claims may be often made. The Treasury might perhaps refer the memorial to the B^d of Trade & Adm^{lty} for a report, this would lay the grounds for giving some compensation for losses if any could be established & then if some bounty could be given to the children now unprovided for, all would be done that could be expected.

It was done, as much as could be expected (but no more); how unfortunate that it could not have been done long before, removing at least one of the problems that so troubled FitzRoy in his later years. A subscription from his friends (with Darwin prominent among them) topped up the grudging contribution from the government (God forbid that they should create a precedent by repaying expenses incurred on government service!), and his widow was provided with accommodation for life in a grace-and-favour apartment, in the gift of the Queen, at Hampton Court, a royal palace on the banks of the Thames. She died there on 29 December 1889.

FitzRoy's reputation as a meteorologist was probably at its lowest ebb in the remaining decades of the nineteenth century, and experienced only a gradual rehabilitation as meteorology was put on a more scientific footing in the twentieth century. By the end of that century, a few professional meteorologists may have been aware of his role as the first head of the Meteorological Office, the inventor of terms such as 'gale warning' and 'weather forecast'; but the vast majority of people who had heard of Robert FitzRoy at all knew of him only as 'Darwin's Captain', and even seafarers listening out for the gale warnings on the broadcasts of the BBC were mostly unaware of who they had to thank for the service that provided those warnings. All that changed at the beginning of February 2002, when the name of the sea area formerly known as Finisterre had to be changed. The name had been used since 1949 in shipping forecasts issued by the British (the BBC broadcasts themselves had begun in 1926) to refer to a large area of sea to the west of the Bay of Biscay; it got its name from the northwestern tip of Spain, dubbed Finisterre by Mediterranean mariners long ago, in reference to its location at (to them) the end of the Earth. But the Spanish use the same name to refer to a smaller patch of sea in roughly the same location, and in a reorganisation carried out by the World Meteorological Organisation the British Finisterre had to go. Overnight, it became sea area Fitzroy – the only sea area in the British system to be named after a person, not a geographical feature. The litany of the BBC shipping forecast has become part of the British way of life since 1926, with names such as Dogger, Humber, Thames, Viking, Forties and Rockall providing a soothing background to everyday life. Now, Fitzroy is part of that litany, becoming familiar indeed to 'generations yet unborn' when he died, and reminding seafarers who still depend on the gale warnings as a matter of life and death just who they have to thank for the service.

Loose Ends

FUEGIANS, *BEAGLE* AND SON

What became of the Fuegians after FitzRoy finally left South American waters? Fate seems to have been kindest to Fuegia Basket. In 1842, when she would still have been in her twenties, the captain of a sealing schooner based in the Falkland Islands reported that on a visit to the western end of the Beagle channel he had encountered a native woman who spoke some English, and that she 'lived some days on board'.[1] Since Fuegia was the only native woman known to speak English, it seems likely that it was her. Three decades later, in 1873 a young missionary called Thomas Bridges definitely encountered Fuegia, by then in her fifties – a great age for a Fuegian, but seemingly happy and in good health. He learned that York Minster had been killed long before by relatives of a man that York had killed (which would have come as no surprise to FitzRoy had he lived to hear the news), but not before she had had at least two sons by him. Ten years later, Bridges met Fuegia again, but by then she was old and sick, and it seems likely that she died soon after.

Jemmy Button, though, had been destined to play a more active (and destructive) part in the establishment of the missionaries in Tierra del Fuego. After several abortive attempts to follow up FitzRoy's lead, in the 1850s a small missionary party set up a base in the Falklands, and deliberately set out to find Jemmy, to act as an intermediary for them. Amazingly, they succeeded – and he was persuaded to take his wife and children to the Falklands for a visit of several months. By now, it was a quarter of a century since he had parted from FitzRoy, and Jemmy must have been in his early forties. Everything seemed to be going well. This first party of Fuegians duly returned to their homeland and a second group of nine natives, including one of Jemmy's grown-up sons, was brought back to the Falklands. But when this party was returned to Woollya, in October 1859, disaster struck. It seems that the natives had been rather thoroughly searched before being allowed to leave the ship, and various items they had stolen were recovered. Whether they were annoyed at the indignity or simply cross at losing their booty we shall never know. But when the Captain and crew of the schooner, the *Allen Gardiner*, were rash enough to go ashore unarmed, leaving only the cook behind, they were slaughtered by the natives. The cook escaped in a dinghy and was eventually rescued alive to tell the tale. Although Jemmy was taken to the Falklands to attend an inquiry into the incident, no punitive action was taken and he was returned to Woollya. The official conclusion of the Governor of the Falklands was that the disaster had been caused by the unwise actions of the missionaries, who had brought the problem upon themselves.

News of the massacre reached England in May 1860, but received only modest attention in the press. Nevertheless, news both of the killings and of Jemmy Button's part in them must have reached FitzRoy about the time of the BA meeting that year, just at the time when he was already upset by the publication of the *Origin* and feeling guilty about his part in providing Darwin with the material on which the theory of natural selection was based. We have no record of FitzRoy's reaction; but the news cannot have been good for his equanimity in the troubled years that lay ahead. Jemmy himself died in 1864, less than a year before FitzRoy, in an epidemic of a measles-like disease brought in by sailors, which killed half the natives – an ironic example of the effects of the Malthusian forces which had also been instrumental in pointing Darwin (and Wallace) towards the theory of natural selection.

But there are happier postscripts to the story of 'the' voyage of the *Beagle*, not least to do with the *Beagle* herself, which was far from finished with voyaging when she was paid off at Woolwich in November 1836. Between 1837 and 1843 she carried out a surveying voyage around the coasts of Australia; it's a sign of how well FitzRoy had looked after the ship that this time the necessary preliminary refit took less than a month, and cost just under £2,400. Initially, this voyage was under the command of John Wickham, FitzRoy's First Lieutenant, and with many of her old officers and men aboard (and, on the voyage out from England, with the then Lieutenant George Grey, later to be FitzRoy's successor as Governor of New Zealand, as a passenger as far as Cape Town, where he chartered a schooner). After Wickham resigned because of ill-health in Australia at the end of 1840 (he later became Governor of Queensland), he was replaced in command by *his* First Lieutenant – John Lort Stokes, who at last received the promotion he so richly deserved. Paid off for the last time on 20 October 1843 (while FitzRoy was en route to New Zealand to take up his post as Governor), in 1845 (the year FitzRoy was recalled from New Zealand) the ship was moved to the River Roach, near Pagglesham, in Essex. There she suffered the indignity of having her masts and sailing gear removed and served as a Coastguard Watch vessel until 1870, although in 1863 her name was removed and she was given the number WV7 (for Watch Vessel 7). On 13 May 1870 (50 years to the month after being launched) she was finally sold for scrap, raising £525, and shortly afterwards broken up at the yard of Murray & Trainer in the Thames estuary.[2]

It's worth saying a little more about the career of Stokes, who spent eighteen years in the *Beagle*, rising from Midshipman to Captain in the same ship (Captain in the sense of being in command of the ship, that is); he was made Post in July 1846, carried out a survey in New Zealand waters between 1847 and 1851 in the paddle steamer *Acheron* (how delighted FitzRoy must have been to hear that news!), and although he retired in 1863 he became Rear Admiral in 1864, Vice Admiral in 1871 and Admiral in 1877 through seniority. He was the last Captain of the *Beagle*, and after she was finally paid off he commented, 'My old friend has extensively contributed to our geographical knowledge,' and described his last farewell:

> I loitered a short time to indulge in those feelings, that naturally arose, in taking a final leave of the poor old 'Beagle', at the same place where I first joined her in 1825. Many events have occurred since my first trip to sea in her. I have seen her under every variety of circumstances, placed in peculiar situations and fearful positions, from nearly the Antarctic to the Tropic, cooled by the frigid clime of South America or parched by the heats of North Australia. Under every vicissitude from the grave to the gay I have struggled along with her and after wandering together for 18 years, a fact unprecedented in the Service, I naturally part from her with regret. Her movements, latterly, have been anxiously watched and the chances are that her ribs will separate and that she will perish in the river where she was first put together.[3]

But, as we have seen, the ship was still to provide service (of a sort) for another quarter of a century before being broken up. There is just one surviving relic of the *Beagle*. Some time in the 1830s, Stokes had a small box made out of wood taken from the ship during the extensive refit which preceded her second surveying voyage to South America. Stokes took the box with him into retirement at Haverfordwest in Pembrokeshire, and in 2003 it was part of an exhibition at the National Maritime Museum in Greenwich.

This still isn't quite the end of our tying up of loose ends. The FitzRoy tradition of public service in general, and naval service in particular, did not die with Darwin's Captain. His son, Robert O'Brien FitzRoy, joined the Royal Navy as a Cadet in January 1853 (a few months short of his fourteenth birthday), became a Lieutenant in September 1859, and was made Post in February 1872. He served with such distinction that in August 1884 he was promoted to Commodore and put in charge of the Training Squadron, flying his flag in the

Active, a ship powered by a combination of steam and sail. He became a Rear Admiral in 1888 (by then, the promotion was not merely on the basis of seniority) and Vice Admiral in 1894, being appointed to the plum job of Commander of the Channel Squadron, where his flagship was the *Royal Sovereign*. His career was rounded off by the award of a knighthood. Between them, the two Robert FitzRoys served in the Royal Navy continuously for more than eighty years, not quite spanning the entire nineteenth century, but covering the transition from the wooden walls of Nelson's Navy to the steam-powered iron battleships of the pre-Dreadnought era. For all his achievements, we suspect that 'our' Robert FitzRoy would have cared more about this continuation of his tradition of service, and the success of his son, than about any of his own achievements.

APPENDIX II

FitzRoy's c.v.

D OCUMENT WRITTEN BY ROBERT FITZROY in March 1852, headed:

PRIVATE AND CONFIDENTIAL

FitzRoy's curriculum vitae, his past service, his health and readiness for future duties

MEMORANDUM

In February – 1818 – Robert Fitz Roy entered the Royal Naval College at Portsmouth (being then 12½), and in October – 1819 – finished the usual three years' course of education, and obtained the first medal.

He was then appointed to the Superb, bearing Sir Thomas Hardy's broad pendant in South America; and sailed to join her, in the Owen Glendower, commanded by the Honble. R. C. Spencer – on board which frigate he remained, – between Brazil and Northern Peru, – until her return to England in 1821: when he joined the Hind, under Captain Rous, served two years in the Mediterranean on board that corvette – (the last four months being with Lord John Churchill), then a short time on board the Cambrian (Captain Hamilton) – and returned to England to pass his examinations.

In July, 1825, he passed in seamanship, before Sir William Hoste and other officers, – (and obtained their marked approbation) – then at the Royal Naval College with twenty-six others, and was placed first. All the questions were worked correctly – many of them by three methods (one being by algebra and spherical trigonometry – the other two practical ways, suitable to quick, or to rigorous calculation, as might be desirable.)

Promotion was immediately awarded by the Admiralty – in consequence of this examination, and the first College Medal. Appointed early in 1825 to the Thetis – Sir John Phillimore – and served in her till that frigate was paid off: – and re-commissioned by Captain Bingham.

Re-appointed immediately to the same ship, and fitted her out, – acting as first lieutenant till manned and ready for sea.

On 1828 – at Rio de Janeiro – Sir Robert Otway (Commander in Chief) proposed to Lieut. Fitz Roy that he should join the Ganges – and soon afterwards made him flag lieutenant.

In November of the same year he was appointed to command the Beagle – surveying ship – from which time, till 1837, he was employed in a long series of hydrographical labour. He was promoted in 1834 by Lord Auckland – but still continued in command of the Beagle – carrying on surveys, and a chain of meridian distances round the globe.

In this vessel, in 1828, the word 'port' was substituted for 'larboard,' both in speaking and writing.

From this beginning, it has since become general: – not only in our Navy, but in the American Service.

In 1835 the position of the Challenger's wreck was found out by Capt. Fitz Roy, who rode some hundred miles in search of it, (partly by night), through the unconquered and dangerous territory of the

Araucanian Indians; then hostile to all white men; – and by him the Blonde (Commodore Mason), was piloted to the place where the crew were saved.

In the course of these voyages – between 1828 and 1836 – more than £3,000. were expended out of Captain Fitz Roy's private fortune, in buying, equipping, and manning, small vessels, as tenders; without which the extensive orders that he had received from the Admiralty could not have been executed within a reasonable time.

This money has not been replaced. It was given, under a perhaps erroneous impression, and no demand for it has yet been made.[1]

In 1837 the Royal Geographical Society honored him with their large gold medal, in acknowledgement of the results of the Beagle's voyages.

In 1838 he was elected an Elder Brother of the Trinity House: in 1841, became Member for Durham; and the next year was appointed to act as Conservator of the Mersey.

In August, 1842, he was selected by the Admiralty to attend on the Arch-Duke Frederick of Austria, in his tour through Great Britain. In March, 1843, he introduced a Bill into Parliament for establishing Mercantile Marine Boards – and enforcing the examination of Masters and Mates in the Merchant Service.

Much of the present Act – called the 'Mercantile Marine Act,' – is taken from that bill.

In April, 1843, Lord Stanley proposed to Captain Fitz Roy to go out to New Zealand as Governor: and he thought it was his duty to undertake the onerous office, however distant and ill remunerated. He gave up his seat in Parliament – and other employments, (though tenable for life by remaining in England), and went, with his family, in a merchant ship, to New Zealand.

His proceedings there gave such umbrage to the New Zealand Company – then very powerful – that they occasioned his recall in 1846.

In 1848 Lord Stanley, in the House of Lords, took occasion to speak at considerable length, and very strongly, in favour of Captain Fitz Roy: and in July, of the same year, Lord Auckland directed him to attend to the new, and then quite experimental, frigate, Arrogant – to be fitted with a screw, and peculiar machinery. From that time, till she

303

was commissioned by him, he continually superintended that ship's arrangements: all of which have answered satisfactorily: though many were quite new and original.

In February, 1850, after having proved the Arrogant in every way – and fairly tired himself out – Captain Fitz Roy was obliged to yield to the effects of fatigue – and anxiety about home affairs – conjoined; which had unnerved him, for a time.

At Lisbon he consulted with Commodore Martin, and gave up his ship, in order that he might settle his domestic affairs, and regain his usual uninterrupted health. A week's change of air only, with absolute rest, sufficed to make him feel himself a different person: and a few months in England, after arranging difficulties that had harassed him, entirely recruited his health, which since that time has been, as it always was throughout his whole previous life – remarkably good.

Active employment was soon sought – and a principal part in the management of the 'General Screw Steam Shipping Company' was accepted. As one of the Managing Directors of that rising and successful Company, he has been busily occupied during more than a year, but has gone out of office by rotation – and does not intend to seek for re-election – as his sole object now is to follow up his own proper profession as soon as he can obtain employment in command of a ship. While engaged with that Steam Company he had opportunities of collecting much information of various kinds that will be useful to his own Profession: the interest of which has ever been uppermost in his mind.

Captain Fitz Roy's health is now perfect. He is free to undertake any service. His age is forty-six. He speaks French, Italian, and Spanish: – has learned Latin and Greek (as dead languages), and has studied various scientific as well as professional subjects.

He is a Fellow of the Royal Society, and of other Societies: – and was elected a Member of the Athenaeum Club last January, without ballot, under the Second Rule.[2]

Gunnery and Steam have been constantly studied by him, and much practised. While at the Trinity House – from 1838 to 1843 – he was frequently in steamers on many parts of the British shores: and during the experimental cruize of Sir Charles Napier's squadron, in 1847, Captain Fitz Roy was a passenger for practice; first on board the Vengeance, and then in the Amphion (screw steamer).

March 15, 1852

APPENDIX III

EXTRACT FROM OBITUARY IN *The Gentleman's Magazine*, June 1865

VICE-ADMIRAL FITZROY

April 30. By his own hand, at his residence, Norwood, Surrey, aged 59, Vice-Admiral Robert FitzRoy, head of the meteorological department of the Board of Trade.

The deceased, who was born July 5, 1805, at Ampton Hall, Suffolk, was the youngest son of General Lord Charles FitzRoy, by his second wife, Frances Anne, eldest daughter of the first Marquis of Londonderry. In February, 1818, he entered the Royal Naval College, Portsmouth, where he was awarded a medal for proficiency in his studies. On October 19, 1819, he was appointed to the 'Owen Glendower,' then coasting between Brazil and Northern Peru. In 1821 he joined the 'Hind,' and served two years in the Mediterranean. At an examination in the Royal Naval College, Portsmouth, in July, 1824, he obtained the first place among twenty-six candidates, and was promoted immediately. In 1825, he joined the 'Thetis,' and in 1828 he was appointed to the 'Ganges,' and soon after flag-lieutenant at Rio de

Janeiro. In November, 1828, Mr. FitzRoy was made commander of the 'Beagle,' a vessel employed in surveying the shores of Patagonia, Terra del Fuego, Chili, and Peru. In the winter of 1829, during an absence of thirty-two days from his ship, in a whale-boat, he explored the Jerome channel, and discovered the Otway and Skyring waters. On December 3, 1834, he was promoted to the rank of captain, but remained in command of the 'Beagle,' pursuing his hydrographical duties, making surveys, and carrying a chain of meridian distances round the globe. During these surveys he expended considerably more than £3,000 out of his private fortune in buying, equipping, and manning small vessels as tenders, to enable him to carry out the orders of the Admiralty, an outlay which was not refunded to him. Captain FitzRoy was elected an elder brother of the Trinity House in 1839, and sat in the House of Commons as member for Durham in 1841. He was appointed acting conservator of the Mersey, September 21, 1842; and in the same year, he was selected to attend the Archduke Frederick of Austria in his tour through Great Britain. He introduced a bill in Parliament in March, 1843, for establishing mercantile marine boards, and enforcing the examination of masters and mates in the merchant service. He went out as governor of New Zealand in April, 1843, and was succeeded in that office by Sir George Grey in 1846. In July, 1848, he superintended the fitting of the 'Arrogant,' with a screw and peculiar machinery which gave the utmost satisfaction. He became rear-admiral in 1857, and vice-admiral in 1863.

When, in 1854, the meteorological department of the Board of Trade was established, Captain FitzRoy was placed at its head, and to him are owing the storm signals and other models of warning that are now in use for the benefit of the seaman. His own life, however, was the price of his devotion to his duties. For some time before his death he had suffered greatly from depression of mind, and had consulted his medical attendant, Dr. Frederick Heatley, who, perceiving that he was much reduced in health by the severe mental labours incident to his position, told him that he must rest from his labours for awhile, and only on the Thursday before his death warned him that he must give up his studies, or the brain would become so affected that paralysis would ensue, but there was nothing in the tone of the Admiral's conversation that could lead to the supposition that he would commit suicide. On the day before his death he called on his friend, Captain Maury, the American navigator, who was about to leave for the West Indies, and his strange condition struck both that officer and a clergyman with whom he was staying. In the afternoon he went to London, returning in the evening. He retired to rest at the usual time, and on the following morning got up earlier than usual, and went to his bath-room. The family, finding that he remained longer than usual, knocked several times on the door, but receiving no answer, the door was at length broken open, when the Admiral was found weltering in his blood, having cut his throat. Dr. Heatley was immediately summoned, and on his arrival the Admiral was alive and recognised him, but he died soon afterwards. These facts having been deposed to by Dr. Heatley and other witnesses, the coroner's jury returned a verdict to the effect that deceased destroyed himself while in an unsound state of mind.

Admiral FitzRoy was a Fellow of the Royal Society, of the Royal Asiatic Society, and many other learned bodies. He published – 'Narrative of the Surveying Voyages of H.M.S. "Adventure" and "Beagle," between the years 1826 and 1833, Describing their Examination of the Southern Shores of South America, and the "Beagle's" Circumnavigation of the Globe,' 4 vols. 8vo.; 'Remarks on New Zealand,' 1846; and 'Sailing Directions for South America,' 1858. He was twice married, first in 1836 to Mary Henrietta, second daughter of the late Major-General O'Brien, which lady died in the spring of 1852; and secondly, in 1854, to Maria Isabella, daughter of the late J. H. Smythe, Esq., of Heath Hall, Yorkshire, who survives him. He leaves a son and two daughters by his first marriage.

Of his personal character and devotion to his duties a distinguished naval officer thus writes: – 'I knew poor dear FitzRoy from his boyhood; a more high-principled officer, a more amiable man, or a person of more useful general attainments never walked a quarter-deck.'

The same issue of the magazine also carried an obituary of Abraham Lincoln.

Sources and Further Reading

Nora Barlow, *Charles Darwin and the Voyage of the Beagle*, Pilot Press, London, 1945.

Nora Barlow, editor, *The Autobiography of Charles Darwin, 1809–1882*, Collins, London, 1958.

Nora Barlow, editor, *Darwin and Henslow: The Growth of an Idea*, University of California Press, Berkeley, 1967.

John Beaglehole, editor, *The Journals of Captain James Cook*, CUP, 1955.

Janet Browne, *Charles Darwin: Voyaging*, Jonathan Cape, London, 1995.

F. H. Burkhardt & S. Smith, editors, *The Complete Correspondence of Charles Darwin*, Volume I, 1821–1836, Cambridge UP, 1985; and subsequent volumes.

Hugh Carleton, *The Life of Henry Williams*, Volume 2, Wilson & Horton, Auckland, 1877.

Christopher Cooke, *FitzRoy's Facts and Failures*, Hall, London, 1867.

Nicholas Courtney, *Gale Force 10: The Life and Legacy of Admiral Beaufort*, Review, London, 2002.

Charles Darwin, *Journal of Researches into the Natural History and Geology of the Countries Visited during the Voyage of H.M.S. 'Beagle' Round the World*, Ward,

Lock & Co, London, eighth edition, 1890 (based on the corrected and enlarged edition of 1845; first edition published in 1839 as volume III of the *Narrative*).

Charles Darwin, *Autobiographies*, edited by Michael Neve and Sharon Messenger, combined edition, Penguin, London, 2002.

Francis Darwin, editor, *Life and Letters of Charles Darwin*, John Murray, London, 1887 (a version of this book under the title *The Autobiography of Charles Darwin and Selected Letters* was published by Dover, New York, in 1958 and is still available).

Francis Darwin and A. C. Seward, editors, *More Letters of Charles Darwin*, John Murray, London, 1903.

L. S. Dawson, *Memoirs of Hydrography*, combined edition, Henry Keay, Eastbourne, 1969 (originally published in two parts, 1885).

John Derry, *Castlereagh*, Allen Lane, London, 1976.

Bernard Falk, *The Royal Fitz Roys: Dukes of Grafton through four centuries*, Hutchinson, London, 1950.

Robert FitzRoy, *Remarks on New Zealand*, White, London, 1846.

Robert FitzRoy, *The Weather Book: A manual of practical meteorology*, Longman, Green, Longman, Roberts, & Green, London, 1862 (second edition, 1863).

Albert Friendly, *Beaufort of the Admiralty*, Hutchinson, London, 1977.

Nick Hazlewood, *Savage: The Life and Times of Jemmy Button*, Hodder, London, 2000.

Wendy Hinde, *Castlereagh*, Collins, London, 1981.

Montgomery Hyde, *The Strange Death of Lord Castlereagh*, Heinemann, London, 1959.

R. Brimley Johnson, editor, *The Novels of Captain Marryat*, collector's edition, Dent, London, 1896. *Mr Midshipman Easy* and *Peter Simple* are particularly relevant.

L. J. Jordanova, *Lamarck*, Oxford UP, 1984.

Richard Keynes, editor, *Charles Darwin's Beagle Diary*, Cambridge UP, 1988.

Richard Keynes, *Fossils, Finches and Feugians*, HarperCollins, London, 2002.

Parker King, Pringle Stokes, Robert FitzRoy & Charles Darwin, *Narrative of the Surveying Voyages of His Majesty's Ships Adventure and Beagle*, Henry Colburn, London, 1839. Published in three volumes. Volume I mostly based on material provided by King and edited by FitzRoy, with contributions from Stokes and FitzRoy; Volume II by FitzRoy (this is also the basis for the version edited by David Stanbury); Volume III by Darwin (later published separately as Darwin's *Journal of Researches*; see also the version edited by Nora Barlow). Referred to in the text as *Narrative*, with the appropriate volume number.[1]

Desmond King-Hele, *Erasmus Darwin*, De La Mare, London, 1999.

Edward Larson, *Evolution's Workshop*, Allen Lane, London, 2001.

H. E. Litchfield, editor, *Emma Darwin: A Century of Family Letters*, Murray, London, 1915 (first printed privately by Cambridge University Press for Litchfield, Darwin's daughter, in 1904).

H. L. McKinney, *Wallace and Natural Selection*, Yale UP, New Haven, 1972.

A. H. McLintock, *Crown Colony Government in New Zealand*, Government Printer, Wellington, 1958.

T. R. Malthus, *An Essay On the Principle of Population*, Johnson, London, 1826 (this is the sixth edition, which Darwin read; also available in a modern Penguin reprint. The first edition appeared anonymously in 1798).

Richard Lee Marks, *Three Men of the Beagle*, Knopf, New York, 1991.

H. E. L. Mellersh, *FitzRoy of the Beagle*, Mason & Lipscomb, New York, 1968.

Paul Moon, *FitzRoy: Governor in Crisis*, David Ling, Auckland, 2000.

Alan Moorehead, *Darwin and the Beagle*, Hamish Hamilton, London, 1969.

Geoffrey Penn, *Snotty: The Story of the Midshipman*, Hollis & Carter, London, 1957.

Samuel Pepys, *The Shorter Pepys*, ed. Robert Latham, Bell & Hyman, London, 1985.

G. S. Ritchie, *The Admiralty Chart: British Naval Hydrography in the Nineteenth Century*, revised edition, Pentland Press, Durham, 1995.

David Stanbury, editor, *A Narrative of The Voyage of H.M.S. Beagle*, Folio Society, London, 1977. Based on extracts from FitzRoy's *Narrative* plus other material.

J. L. Stokes, *Discoveries in Australia*, Boone, London, 1846 (in two volumes).

Henry Sulivan, *Life and Letters of the late Admiral Sir Bartholomew James Sulivan, K.C.B.*, John Murray, London, 1896.

Arthur Thomson, *The Story of New Zealand*, John Murray, London, 1859 (in two volumes).

Keith Thomson, *HMS Beagle*, Norton, New York, 1995.

E. J. Wakefield, *Adventure in New Zealand*, John Murray, London, 1845; in two volumes.

F. Walker, *Young Gentlemen: The Story of Midshipmen*, Longman, Green & Co, London, 1938.

Geoffrey West, *Charles Darwin: The Fragmentary Man*, Routledge, London, 1937.

Ormond Wilson, *Kororareka and Other Essays*, John McIndoe, Dunedin, 1990.

Endnotes

Introduction

[1] In the early part of his career, that name is sometimes given in official documents as Fitz Roy, and in some of his own early writings, Robert refers to himself as Fitz-Roy. But FitzRoy became the usual form long before he died, and we shall use it throughout this book, for consistency.

Chapter One

[1] Legend has it that on his restoration in 1660, Charles spent the first night at his palace in Whitehall in the arms of the nineteen-year-old Barbara, who had married Roger Palmer the previous year. Her first child, a daughter, was born exactly nine months after the Restoration. Palmer, although far from happy with the situation, became Earl of Castlemaine for his services to the King in 1661, and Barbara left him in 1662.

[2] See Falk. Full references to works cited are given in the Bibliography.

[3] Although George I was, through his mother, a great-grandson of James II, as grandson of Charles II in a direct male line the second Duke was, in a

sense, more 'royal' than his King. But by 1714, when George came to the throne, the English had stopped fighting wars over such matters.

4 Falk.

5 This career path would not be possible today, since the Prime Minister is now also the First Lord of the Treasury.

6 The ex-Duchess moved even more swiftly, marrying within twenty-four hours of the passage of the Bill for the divorce through the House of Lords, the only way a member of that House could obtain a divorce. Falk reports the delicious story that she began a letter to a friend the day before the Bill passed, and signed it 'Anne Grafton'. The next day, she added a little to the letter, and signed the addition 'Anne Liddell'. Before the letter was sent off, she had remarried, to the Earl of Upper Ossory, and added a bit more to the letter, signing it 'Anne Ossory'. The story soon became common knowledge, and provoked the contemporary epigram:

No grace but Grafton's grace so soon
So strangely could convert a sinner,
Duchess at morn, and miss at noon
And Upper Ossory after dinner.

7 Even here, though, he was overshadowed by his father. The third Duke owned three horses that won the Derby: Tyrant in 1802, Pope in 1809 and Whalebone in 1810.

8 For the record, Castlereagh had married Emily Hobart in 1794, and the couple seem to have been devoted to one another, although they had no children. The unconfirmed story which Castlereagh told his closest friends was that he had been lured into a brothel by an attractive young 'woman' who turned out to be a man, and that accomplices of the transvestite had confronted the pair in a compromising situation and made the blackmail threats. But this is just one of many stories Castlereagh told at the time, most of which were clearly (to others) paranoid delusions.

9 Quoted by Hinde.

10 1 June 1866.

11 Some are preserved in the archive at the University of Cambridge Library.

12 That is, moved up.

13 See Bibliography.

14 See also the books by Geoffrey Penn and C. F. Walker cited in the Bibliography.

15 Both examples quoted by Penn.

16 James's mother had studied music under Mozart when living in London.

17 Such officers were also known as Mates, but there was another quite different rank of Mate (or Master's Mate) in the Royal Navy as well, and we don't want to confuse the issue so we shall stick with Midshipman.

18 See Bibliography.

19 Sulivan.

20 He was originally appointed to the *Superb*, in South American waters, and was travelling on the *Owen Glendower* to take up that posting; but as he makes clear in his c.v. (see Appendix II), for some unexplained reason he

stayed with the *Owen Glendower* and never actually joined the *Superb*.

21 Quoted by Penn.

22 Quoted by Penn.

23 *Charting the Sea in Peace and War*, HMSO, London, 1947.

24 A fourth collegian later joined the *Thetis* from another ship, the *Seringapatam*, during Sulivan's time with her.

25 A barque is a small sailing vessel with at least three masts, with the foremasts square-rigged and the rear mast carrying a fore-and-aft rig. More about the *Beagle* herself in the next chapter. Although Sulivan is perfectly correct in referring to the *Beagle* as 'H.M. barque', or 'H.M.B.', it is also correct to refer to her as HMS *Beagle*.

26 The famous occasion when Nelson clapped his telescope to his blind eye in order to ignore a signal from his Admiral ordering him to call off the action.

27 In the Royal Navy, the officer in command of a ship is always given the courtesy title of Captain, whatever his substantive rank, so we shall refer to Captain FitzRoy, not Commander FitzRoy, when describing his time on board the *Beagle*.

28 This time Sulivan uses another technically correct but confusing naval term; in the Royal Navy of the time, the term sloop was often used generically to refer to any small vessel armed with fewer than twenty guns.

Chapter Two

1 See *Narrative*, Volume I; this is the main (almost the only) source of information about FitzRoy's first voyage with the *Beagle*, and all quotations in this chapter are from that source unless otherwise cited.

2 A kind of sleeping bag made from a blanket sewn up at the sides and with a drawstring around the neck.

3 Sulivan's autobiographical note says 28.5 inches, but this was written long after the event and King's account is based on the ship's log of the *Adventure*.

4 At 330 tons, *Adventure* was significantly bigger than the *Beagle*, and described by FitzRoy as 'roomy'. Roomy, that is, if you were used to the *Beagle*.

5 A slight piece of literary licence; King wasn't an admiral. But according to Sulivan they had indeed just made out the *Adventure*, 'seeing her higher spars over Goritte Island, off Maldonado', when the storm struck. He describes the pampero as the heaviest any vessel ever experienced and survived.

6 This is the classic schooner rig used on the *Adelaide*; larger schooners with more masts were also built.

7 Which is actually a network of channels between the islands of the tip of South America. Patagonia is to the north of the Strait, while to the south the largest island, which forms most of the tip of the continent, is Tierra del Fuego. FitzRoy uses the term 'Fuegians' to refer to natives from either side of the Strait.

8 FitzRoy mentions one man, 'a careful north countryman', who 'carried with

him, when he left the Beagle, two new pair of shoes (besides those on his feet), and three pair of new stockings: but brought back only a ragged pair of stockings and the remains of one shoe. The others had been fairly worn out, or lost, in scrambling over rocks and ascending mountains.'

9   Chiloé had, incidentally, until recently been the last possession of the Spanish Crown in South America.

10 From that description, it sounds like a crudely made coracle; but FitzRoy later makes it clear that the curious craft was elongated, like a canoe, not circular, like a coracle.

11 The long line, weighted with lead, used to take soundings from the boat.

12 He left the children with an old woman he met at the westernmost part of his survey, 'who appeared to know them very well, and to be very much pleased at having them placed in her care'.

13 As they soon would be; she came to the throne in 1837, but the patronising attitude of the British towards 'natives' was 'Victorian' even before that.

14 To little avail; the King died on 26 June 1830.

Chapter Three

1   As a key member of the crew, a coxswain often accompanied the same captain from ship to ship, and attended him ashore between ships.

2   The woman was Mary Henrietta O'Brien, the daughter of an Irish country gentleman and retired Army Major General, a widower; FitzRoy seems to have met her and formed an understanding during the time he spent in England between his two voyages in command of the Beagle, but almost nothing is known about her prior to their marriage shortly after he returned from the second Beagle voyage.

3   An officer was said to be 'made Post' when he was officially promoted to the substantive rank of Captain, with the news being posted in the records of the Navy; this distinguished a Post Captain from an officer with lesser rank who received the courtesy title of Captain while in command of a ship.

4   Reproduced in one of the appendices to the Narrative.

5   Quoted in the Good Words obituary of FitzRoy.

6   The youngster had actually first gone to sea at the age of nine, in his father's ship. He would leave the Beagle in Australia to rejoin his father.

7   This refers to Erasmus Darwin, a successful doctor who was also widely known for his poetry in the late eighteenth century, and a natural philosopher who wrote about evolution, although it is not clear whether these writings influenced his grandson.

Chapter Four

1   Quoted by Francis Darwin, 'FitzRoy and Darwin, 1831–36', Nature, Volume 88, pages 547–8, 1912.

2   Wildwood House, London, 1974.

3   Josiah Wedgwood II, Robert Darwin's brother-in-law.

⁴ *Cornhill Magazine*, April 1932.
⁵ John Bowlby, *Charles Darwin*, Norton, New York, 1990.
⁶ A well-known ascetic.
⁷ Things were rather different away from the Captain's table. Sulivan describes the rations allowed on one of the small-boat surveying expeditions as 'only two pounds of meat, two-thirds of a gallon of tea, one pound of bread, and a quarter of a pint of rum each man per day'.

Chapter Five

¹ There were, of course, sober *officers*.
² Quotes from FitzRoy in this chapter are from Volume II of the *Narrative* unless otherwise specified. Beccaria is Cesare Beccaria, an eighteenth-century Italian philanthropist who was an opponent of capital punishment and coined the phrase 'the greatest happiness of the greatest number'.
³ The diary provided the raw material on which the published *Journal* (Volume III of the *Narrative*) was based, but has now been published in its own right. This is why you sometimes see quotations from Darwin describing the same events in slightly different words. In FitzRoy's case, the original notes have been lost, and we only have the published version of his contribution to the *Narrative*.
⁴ See *Voyaging*.
⁵ *Complete Correspondence*.
⁶ *Autobiographies*.
⁷ That is, a survey ship.
⁸ Except for the surgeon, Robert McCormick. It was traditional in Royal Navy ships for the surgeon, the nearest thing to a scientist on board, to act as naturalist, and McCormick had hoped to gain a reputation in this field from the *Beagle* voyage. Darwin's presence, and FitzRoy's friendship with Darwin, made it clear that this would never happen, and soon after reaching Rio he had obtained permission to return home, leaving his former assistant, Benjamin Bynoe, as Acting Surgeon.
⁹ *Beagle* carried nine guns on this second voyage, one of them in the bow, so a 'broadside' amounted to four tiny pop-guns!
¹⁰ That is, at the expense of.
¹¹ The fore-and-aft schooner rig was ideal for this, as well as enabling such ships to beat effectively to windward.
¹² Letter in Hydrographic Department archive.
¹³ We have to find the details from Sulivan, who wrote in a letter home: 'I dived down under the keel, and, having ascertained things were not so bad, came up the other side, bleeding from several scratches received from the jagged copper. Captain FitzRoy, wishing to make doubly sure, then performed the same action himself.'
¹⁴ See Appendix I.
¹⁵ FitzRoy's description of the islands in a letter to Beaufort reads: 'In the appearance of the Falkland Islands, there is very little either remarkable or

interesting. Barren hills, of moderate height, sloping gradually towards a low and rugged coast, characterise the greater part of these Islands.'

[16] FitzRoy is not expressing his disappointment with any of these people, but his disappointment that his recommendations that they be promoted have not been acted upon; as ever, he is as upset about what he sees as unjust treatment of his subordinates as he is about the unjust treatment to himself.

[17] It's ironic that Darwin, who had been taken on the voyage in no small measure to help the Captain keep his sanity, should have been ill just at the time when he might have done most good in this regard.

[18] The stay in Valdivia was also remembered by Darwin for 'an unusual degree of gaiety on board. The *intendente* paid us a visit one day and brought a whole boat full of ladies: bad weather compelled them to stay all night, a sore plague both to us and them. They in return gave a ball, which was attended by nearly all on board. Those who went returned exceedingly well pleased with the people of Valdivia. The signoritas are pronounced very charming; and what is still more surprising, they have not forgotten how to blush, an art which is at present quite unknown in Chiloe.'

[19] The promotion actually dated from 3 December 1834, but the news had only just reached Valparaiso.

[20] They even sold the little ship for a little more than FitzRoy had paid for her – but not enough to cover all his expenses including her outfitting.

[21] We shall go into the background history of New Zealand in Chapter Seven.

Chapter Six

[1] Published in the *Journal of the Royal Geographical Society*, Volume 6, pages 311–43, 1836.

[2] *The Life-Boat*, 2 October 1865.

[3] So he must have been at least ten years at sea by then.

[4] FitzRoy's steward.

[5] The Second Secretary at the Admiralty.

[6] FitzRoy's brother-in-law and his cousin.

[7] There is just one genuine joint publication by FitzRoy and Darwin – an article they wrote together on the voyage home about the work of the missionaries in New Zealand, and which was published in the *South African Christian Recorder* in Cape Town, through their contacts with the astronomer John Herschel, who was based there at the time ('A letter containing remarks on the Moral State of Tahiti, New Zealand, etc.', *South African Christian Recorder*, September 1836, pages 221–38). The main interest of this is that it shows how much in tune with one another they still were even at this late stage of the circumnavigation.

[8] *Cornhill Magazine*.

[9] Both Darwin and FitzRoy would have been astonished if they had known that in 2002 a complete set of all four volumes of the *Narrative*, admittedly in good condition, would change hands for £22,500.

[10] See Stanbury.

[11] Their other children were Robert (born 2 April 1839), Bertram (born 11 September 1840, died 4 June 1841), Fanny (27 April 1842) and Katharine (22 March 1845). The two girls lived well into the twentieth century, Fanny dying in 1922 and Katharine in 1927; Robert died in 1896.

[12] See page 156.

[13] This was a time of personal sadness, however. Their second son, Bertram, who had been born in September 1840, died in June 1841. As always, FitzRoy's response to personal grief was to fling himself into his work.

[14] *Captain FitzRoy's Statement re Collision between William Sheppard and the Author* (W. White, London, 1841) and *The Conduct of Captain Robert FitzRoy R.N. in reference to the Electors of Durham and the Laws of Honour, exposed by William Sheppard Esquire* (John Ollivier, London, 1842).

[15] All of this shows how much duelling had gone out of fashion; things would have been handled much more correctly a generation or two earlier!

[16] Contrast him with FitzRoy at the same age, twenty-six, taking the *Beagle* on her second voyage to South America.

[17] See *Hansard*.

[18] Gladstone was at this time a Tory, but later achieved fame as a Liberal Prime Minister.

Chapter Seven

[1] At that time still a Lieutenant, but Captain of the barque *Endeavour*.

[2] Quoted by McLintock. Other quotes in this chapter from the same source unless otherwise specified.

[3] J. E. FitzGerald, in New Zealand. See *Dictionary of New Zealand Biography* at http://www.dnzb.govt.nz/DNZB/

[4] Letter to Charles Torlesse, quoted by McLintock.

[5] You might wonder why any of the chiefs should have signed away 'ownership' of New Zealand; this phrase probably holds the clue, since they were promised that the British would bring the rule of law, protecting them from the colonists and ensuring their freedom to lead their own way of life on their own lands. In other words the usual sad but familiar tale of how Europeans treated 'uncivilised' people.

[6] Though it would have been news to the vast majority of the 100,000 or more Maoris who lived there.

[7] Only in the sense that he had got better rather than worse. Contemporary accounts say that his illness was caused by the first of a series of strokes which left him unfit for his post for the few years that remained of his life; if so, he should have handed over to Bunbury and retired to New South Wales at once, both for his own good and that of New Zealand. The reason he didn't may be explained by the view of modern historians, such as Paul Moon, that his death was a result of mercury poisoning from medication taken to treat syphilis.

[8] One reason for this was the priority being given to India at the time.

[9] *Remarks on New Zealand*; see below.

10 To modern eyes, it may seem odd that such posts should go to relatively junior naval officers, but as G. W. Hope put it in a speech to the House of Commons in 1845 (see *Hansard*, 18 June 1845): 'naval officers are accustomed to command men from the age of thirteen. In all infant Colonies they selected a naval man to cut through the jungles; when the Colony prospered, a military man superseded the naval man; and, finally, when revenue improved and macadamised roads were formed, the civilian replaced the soldier.'

11 That is, the senior Civil Servant who ran the department under whatever politician happened to be Secretary of State. Stephen was such a powerful figure in the Colonial Office that he was sometimes referred to, behind his back, as the 'Over Secretary'; a modern version of the relationship between a Minister and his Permanent Under-Secretary has been satirised in the TV series *Yes, Minister*.

12 Thomson. His book provides a largely eye-witness account of the early colonial period in New Zealand history; he served there as Surgeon Major with a detachment of the 58th Regiment of Foot, sent over from New South Wales in 1845. The account of FitzRoy's arrival is secondhand, but told directly to Thomson by those who were present.

13 Though even there this had caused problems, notably, as we have seen, with his purchase of auxiliary ships to speed up the surveying work.

14 *Adventure in New Zealand*.

15 Jerningham Wakefield, quoting from an eye-witness account.

16 Thomson mentions that there were just twelve Europeans and 500 natives present on this occasion.

17 FitzRoy's *Remarks*, published in 1846, soon after his return to England, is not the kind of bitter polemic that might be expected from a lesser man in his position, but offers a calm, thoughtful appraisal of the situation, defending his decisions and pointing out, without rancour, the flaws in British policy towards New Zealand.

18 See McLintock.

19 In his *Remarks*, FitzRoy quotes the instance of natives who promptly converted from Protestantism to Catholicism when a French bishop arrived in New Zealand in 1838. When asked why they did so, they replied, 'We like to try what is new.'

20 Some of Mary FitzRoy's letters are housed in the Mitchell Library in Sydney; this also holds correspondence of Philip Parker King. There are others in the Dixson Library in New South Wales. In those days of slow communications and anxieties about health and other dangers, she always writes at the top of the first page of a letter 'all well' or 'all quite well' to reassure the recipients that there are no bad tidings to come.

21 *Remarks*. Other quotes from FitzRoy are from the same source.

22 We should say that technically FitzRoy ruled New Zealand with the aid of a Legislative Council; in practice, the Council more or less did what he told them to do.

23 2 November 1844.

[24] At almost the same time, in a letter dated 16 September 1844, Mary FitzRoy tells her sister-in-law Fanny that 'the condition of the colony is such as to cause the deepest anxiety', that there is 'absolutely no money' other than 'paper shillings & sixpences' and that 'all Gov^t. Officer's are on half pay'. She hopes that 'in England some steps may be taken for our [that is, the Colony's] relief'.

[25] See McLintock.

[26] No relation! Earl Grey became Colonial Secretary when the Liberals came to power in the summer of 1846; from November 1845 to July 1846, Stanley's immediate successor, in the last months of Peel's Tory administration, was William Gladstone, later a Liberal Prime Minister.

[27] Thomson points out that the long-term importance of this alliance has often been underestimated. It provided a precedent making it acceptable for other chiefs, who 'in all important affairs are guided by ancient usages', to side with the Europeans in future conflicts.

[28] According to Thomson, things became so desperate in the south that at Nelson 'seed-potatoes planted in the ground were actually dug up for food'.

[29] The birth was so difficult, and Mary so ill afterwards, that she did not write to Fanny about it until 11 June. In her own words, for several weeks her life 'hung upon a thread', just at a time of political crisis for FitzRoy. General O'Brien seems to have been a domestic rock on this occasion.

[30] Correspondence in the Public Record Office, London.

[31] See *Hansard*.

[32] Trelawney Saunders; see below.

[33] Captain Rous, under whom FitzRoy had served as a Midshipman, and who was now an MP, questioned the reasons given for his dismissal, and said that 'an angel from heaven could not reconcile the differences between the natives, the missionaries, and the New Zealand Company'. See *Hansard*.

[34] 'Governor Grey,' FitzRoy reports without embellishment in his *Remarks*, 'brought money and additional forces, both military and naval.' If FitzRoy had had the money and resources, there would have been no need for Governor Grey.

[35] The first in the world to grant women the vote, in 1893, and a pioneering provider of old-age pensions, in 1898.

[36] Ian Wards, *Dictionary of New Zealand Biography*, http://www.dnzb.govt.nz/DNZB/

[37] See *The New Zealander*, 31 December 1853.

[38] See Appendix I.

[39] Public Record Office.

Chapter Eight

[1] There would have been his half-pay from the Navy while he was still on the Active List though without a ship or other official employment, but hardly enough to maintain the family in an appropriate manner.

[2] See Appendix II.

³ This was just as true in 1834, and shows how close FitzRoy came to seeing his naval career ended at that time.

⁴ Although he remained professionally active and officially available for employment. Among other things he was deeply interested in proposals to cut a canal through the Isthmus of Panama, a topic he spoke on at the Royal Geographical Society; and at Beaufort's instigation he was considered as a candidate for a seagoing command to carry out a tidal survey, but this came to nothing.

⁵ In other words, Robert FitzRoy's Stewart grandparents were also the great-great-grandparents of Winston Churchill.

⁶ Some of the correspondence is in the Cambridge archive.

⁷ In 1853, there was further disappointment when he applied unsuccessfully for the post of Superintendent of the Compass Department.

⁸ It's also interesting that around this time he seems to have changed his political views, and become a supporter of the Liberals (judging by a letter dated 7 November 1853 sent to Norton Shaw, the Secretary of the Royal Geographical Society). But we can only speculate to what extent all these personal troubles contributed to that change of heart.

⁹ See *Good Words*.

Chapter Nine

¹ We now know that the winds *spiral* into the centre of low pressure, so both ideas were partly right.

² The early history of the Meteorological Office, including FitzRoy's contribution, has been summarised by Jim Burton in *British Journal for the History of Science*, Volume 19, pages 147–76, 1986, whose approach we follow in this chapter.

³ Both terms are used in documents from the time; in his *Weather Book*, FitzRoy describes himself as head of the Office for Meteorologic Objects (that is, objectives). We will use the term Office, both to make it clear that this was then a sub-department of the Marine Department, and to match the modern name, Meteorological Office, which FitzRoy also used (among others) in his correspondence.

⁴ The present broad avenue of Parliament Street was at that time two narrow streets, parallel to one another with a row of buildings between them; the Meteorological Department's offices were in that building, now demolished.

⁵ The FitzRoy letter is in the Royal Society Library; Hooker's letter is in Cambridge University Library, and has also been reprinted in, for example, L. Huxley, *Life and Letters of Sir Joseph Dalton Hooker*, Murray, London, 1918.

⁶ Figures in this chapter for the work of the Meteorological Office come from the *Annual Reports* submitted by FitzRoy, and from Burton.

⁷ But he is best known as one of the two people (the other was John Couch Adams, in England) who independently predicted the existence and location of the planet Neptune by studying perturbations in the orbits of the planets.

The discovery had been made in 1846, when le Verrier was thirty-five years old. Le Verrier also introduced into meteorology the idea of plotting on a map lines joining places of equal atmospheric pressure, so-called isobars, the meteorological equivalent of contour lines.

[8] The flow results from a combination of the effects of convection and the rotation of the Earth, and, of course, at other places it goes the other way, closing the loop of atmospheric circulation. In his book, FitzRoy notes that his studies of synoptic charts show that, superimposed on the swirling winds associated with depressions and so on, 'the entire mass of the atmosphere in our latitude, has a constant, a perennial movement toward the east'.

[9] And which will be familiar to readers of Anthony Trollope!

[10] Reported later, in *The Life-Boat*, 2 October 1865.

[11] Lord Farrer, in a letter quoted by Henry Sulivan. A key factor in Sulivan's appointment was that he was owed some reward for distinguished service in the Baltic during the Crimean War; it was this, rather than any failing on FitzRoy's part, that tipped the balance in his favour.

[12] This was the meeting now famous for the fierce debate about Darwin's *Origin of Species*, which had been published in 1859; FitzRoy was deeply affected by the controversy, and we shall return to it shortly.

[13] At Aberdeen, Berwick, Hull, Yarmouth, Dover, Portsmouth, Jersey, Plymouth, Penzance, Cork, Galway, Londonderry and Greenock.

[14] The pen-name came from the Latin proverb *Nemo Senex metuit louem*, which translates as 'An old man should be fearful of God.' FitzRoy was only fifty-four, but the choice of name suggests that he was beginning to feel his age.

[15] A comment made by him at the BA meeting in 1860.

[16] Letter now in the possession of Simon Keynes.

[17] No official transcription was made of the debate, and different sources quote slightly different versions of the exact words used by the protagonists. See, for example, the *Life and Letters* volumes of both Huxley and Joseph Hooker. The FitzRoy quote (or possibly paraphrase) comes from the *Athenaeum*, containing the official record of the meeting. But FitzRoy's intervention clearly made an impact. In a letter to Darwin dated 15 November 1866, the biologist Julius Carus said, 'I shall never forget that meeting of the combined sections of the British Association when at Oxford in 1860, where Admiral FitzRoy expressed his sorrows for having given you the opportunities of collecting facts for such a shocking theory as yours.'

[18] In addition, about this time FitzRoy must have become aware of the latest unhappy circumstances in the attempts to establish a mission station in Tierra del Fuego (see Appendix I).

[19] See *The Weather Book*.

[20] See Burton.

[21] See Mellersh.

[22] See D. Barlow, *Weather*, Volume 52, Number 11, page 337, November 1997.

23 Not that this was a particularly new problem; back in January 1858 FitzRoy had sent a typical letter to Williams, justifying his unorthodox attitude to accounting procedures by saying that 'where much work is to be done – and done quickly – strict adherence to forms may be sometimes prejudicial, while the object of all routine – namely – good work – is to be sacrificed'. By their standards, the Civil Service really did bend over backwards to try to accommodate what they saw as FitzRoy's foibles!

24 It's interesting that FitzRoy's handwriting is smaller and neater in letters written from the *Beagle* in the old days. The change seems to have happened in New Zealand at the time he was Governor; but we leave it for psychologists to discuss any possible significance in this, assuming it wasn't simply related to deteriorating eyesight!

25 See Mellersh.

26 Coincidentally, Lincoln had been born on the same day as Charles Darwin, 12 February 1809.

27 Not really 'girls' – Fanny was twenty-three the year her father died, and Katherine twenty. Laura was just seven, and twenty-six-year-old Robert O'Brien FitzRoy was already well established in his own career in the Royal Navy (see Appendix I).

Chapter Ten

1 *The Life-Boat*, 1 July 1865.

2 *Good Words*.

3 See Mellersh.

4 Cited in the Sulivan biography. First as plain Thomas Henry Farrer (he did not become a baron until 1893), and from 1883 as Sir Thomas, he served in the marine department of the Board of Trade and then as Permanent Secretary from 1850 to 1886; his daughter Ida married Charles Darwin's youngest son, Horace, in 1880. Sulivan is the other obvious candidate as author of the testimonial in *The Life-Boat*, if it was not Stokes.

5 Then Chancellor of the Exchequer.

6 Presumably, Edward O'Brien had left a small inheritance for his grandchildren, but FitzRoy's widow and his youngest daughter Laura had nothing of their own.

Appendix I

1 See Marks.

2 Some accounts say that the ship was sold to Japan and ended her life in the far east; but this was a different vessel with the same name, a paddle steamer.

3 Quoted by Keith Thomson.

Appendix II

[1] Other smaller sums were repaid; amounting to about £700: but these were exclusive of the above-mentioned £3,000.
[2] An honorary mode of election.

Sources and Further Reading

[1] Just as we were completing this book, we learned that a complete facsimile edition of all four volumes was published by AMS Press, New York, 1966. Unfortunately, it is out of print.

Index

More Non-fiction from Headline

THE FLOATING BROTHEL

SIÂN REES

'*Not much attempt had been made to enforce discipline among the women, many of them London prostitutes, who had turned the ship into a floating brothel at her various ports of call.*'

In July 1789, 237 women convicts left England for Botany Bay in Australia on board a ship called the *Lady Julian*, destined to provide sexual services and a breeding bank for the men already there. This is the enthralling story of the women and their voyage.

Based on painstaking research and primary sources, such as court records and the first-hand account of the voyage written by the ship's steward who fell in love with 19-year-old Sarah Whitelam, this is a riveting work of recovered history.

'A miracle of compressed intelligence' *The Times*

'Simultaneously shocking and utterly engrossing . . . [a] wonderfully vivid book, beautifully written and all the more enthralling for being historically accurate' *Daily Mail*

NON-FICTION / HISTORY 0 7472 6632 8

MEASURING THE UNIVERSE

KITTY FERGUSON

An exciting and accessible mix of history and cutting-edge science, *Measuring the Universe* chronicles man's attempts to push back the frontiers of understanding.

How far . . . how fast . . . how big . . . how old . . . who are we? The questions range from the mechanical to the metaphysical, but all are rooted in centuries-old curiosity, and much besides mathematics and astronomy has gone into answering them. Politics, religion, philosophy and personal ambition have all played roles in this drama.

Measuring the Universe examines these scientific puzzles and the tricky human milieu in which they have been studied, argued, and sometimes solved. From the early Greeks' experiments in working out the circumference of the earth, to the latest probings beyond the borders of the observable universe, intelligent minds have asked questions about our universe, and our place in it. This compelling adventure story whisks us through three millennia of scientific ingenuity and intellectual history, and leaves us poised on the twenty-first century cutting edge of science.

'It is one of the great stories of science and Ferguson tells it well' *Sunday Telegraph*

'She turns men of science into men of fascination' *Evening Standard*

NON-FICTION / POPULAR SCIENCE 0 7472 5699 3

ECLIPSE
The Celestial Phenomenon which has Changed the Course of History

DUNCAN STEEL

Since the dawn of time, eclipses have been perceived as peculiarly portentous events. Whether they were seen as signs of divine displeasure or augurs of good fortune, a great deal was at stake when it came to predicting when next they might occur. Stonehenge may well have been built for this very purpose and all ancient civilisations set great store by keeping accurate records of these celestial shadows.

Such extraordinary phenomena hold a powerful fascination for us all even today, and here Duncan Steel explains everything you will ever need to know about eclipses, their science and their significance to humankind. This fascinating and epic story – peopled by characters such as Julius Caesar, William Shakespeare, Christopher Columbus and Albert Einstein – stretches from ancient Babylon to the present day and beyond . . .

'Everything you need to know on the subject'
The Good Book Guide

NON-FICTION / POPULAR SCIENCE 0 7472 6284 5

Now you can buy any of these other bestselling non-fiction titles from your bookshop or *direct from the publisher*.